CW00539089

THE METEORITE HUNTERS

THE METEORITE HUNTERS

On the Trail of Extraterrestrial Treasures and the Secrets Inside Them

JOSHUA HOWGEGO

A Oneworld Book

First published by Oneworld Publications Ltd, 2025

A CIP record for this title is available from the British Library

ISBN 978-0-86154-919-1
eISBN 978-0-86154-920-7

Typeset by Tetragon, London
Printed and bound in Great Britain by Clays Ltd, Elcograf S.p.A.

Oneworld Publications Ltd
10 Bloomsbury Street
London WC1B 3SR
England

For Jonty and Ewan – may you find joy
and wonder in life, wherever it takes you.

Contents

Author's Note

There is a spot next to a fishpond where Winston Churchill liked to sit and think. I can't describe exactly what it was like in the prime minister's heyday, but today the water is surrounded by lush ferns, magnolia trees, a huge cedar and several patches of bamboo that rustle gently in the breeze. It sits in the landscaped gardens of Churchill's red-brick country house, Chartwell, now open to the public. I'm afraid the tranquillity was shattered by my two boisterous boys when I took them to visit one day, but even they were content to pause for a moment by the pond and watch the fish swim in a dreamy kaleidoscope.

I'm telling you all this because these snatched moments by the pond turned out to have an unexpected connection to meteorites. As all parents know, you sometimes have surprisingly philosophical conversations with your children, and somehow we got talking about whether the fish knew that there is a wide world above the surface of their pond. Presumably they perceive some other blurry reality out there, but they surely can't know the full extent. Then something else caught our eye. For years, visitors to Chartwell have been throwing coins into this fishpond, so that flashes of silver and bronze glimmer around the edges. What did the fish make of these objects, these clues to human civilisation?

This happened a few weeks after I had started working on this book, and it struck me that these pennies were the perfect way to explain to my children what so intrigued me about meteorites. In a sense, we are just like the fish in that pond: imprisoned on the Earth with only a limited view of the expanse of the cosmos beyond. We can look out at it with telescopes, for sure, but we get an imperfect view. Yet we do occasionally get objects delivered to us from somewhere above and beyond the realm we inhabit, like those glinting coins cast into the water.

Meteorites have fascinated people for centuries. We all know that the biggest ones, asteroids, have a frightening, life-smiting power. The impact that ended the reign of the dinosaurs is a case in point. But they also hold a deep cultural and even religious grip on us. For the ancient Egyptians, meteorites were gifts from the gods, associated with divine power. The ancient Greeks wrote of stones that fell from the sky and incorporated them into statues they worshipped. The Clackamas people, a group of indigenous Americans, venerated the huge and bizarrely twisted bulk of the Willamette meteorite in Oregon. Hundreds of years ago, Clackamas warriors are said to have ceremonially dipped their arrows in rainwater that collected in the meteorite's crevices. The fiery trails left by meteors were also seen as portents – of good and ill – for much of recorded history. There's a sense that meteorites connect us with the wider cosmos, an expanse both spatially and temporally far beyond us, but of which we remain a part. C. S. Lewis conjured this up masterfully in his poem *The Meteorite*, part of which is reproduced on page XV. He used the poem as an introduction for his book *Miracles* – and even for the most hardened empiricist, there is something miraculous about a shard of the heavens descending to us.

For my part, though, I had never dwelt much on meteorites until I had them forced upon me during the course of my work as an editor for *New Scientist* magazine. I had been dispatched to another old country house, this time one in Bedfordshire, where a group of scientists were gathering for a conference. My job was to come up with ideas for features for the magazine, and I'd noticed a set of discussions about the origins of water in the solar system. It sounded suitably grand and interesting, so I took an early train from London to see what it was all about.

As I sat through the talks and tried to corner the scientists during the coffee breaks, I realised that nearly all of the discussion centred on meteorites. That was a surprise at first. But as I listened, I started to see that if you want to ask fundamental questions about the Earth, the solar system and why they are as they are, one of the best sources of information turns out to be meteorites. They serve as time capsules that contain the materials from which everything in the sun's orbit was ultimately made – the planets, comets, asteroids and even life itself.

When I got back home to London, I started thinking about meteorites a lot. Although it was simple enough to read up on the basic facts, I found I had deeper questions not so easily answered. Just how does a person even start to think about finding one of these things? And most intriguing of all, how could you read and understand the information encoded within them? Over the next few years, I talked to as many people in the field of meteoritics as I could and started asking them about their world. I found that it was a wild and delightful one, crammed with unlikely characters, adventure and discovery. So I decided to dive in and observe everything so that I could write the stories down. It was a journey that took me to dusty museum basements, the rooftops of Oslo and deep into the wilderness of central Sweden – though that was nothing compared to the expeditions of some of the meteorite hunters I met.

This book isn't supposed to be a comprehensive study guide to meteorites, not least because accomplished works of that sort already exist. Instead, I wanted to keep the aperture focused on two particular subjects. In the first part of the book, I look at the human stories of those who chase down meteorites in all manner of surprising and exciting ways. The past few years have seen a revolution in meteorite-hunting methods, especially because we can now find them in urban environments for the first time and because we have an ever-expanding technological capacity to detect them as they enter the atmosphere. In the second half of the book, I investigate how scientists tease information from inside meteorites and what that tells us. This work is not only helping to drastically rewrite the history of the solar system, but it also sheds light on the puzzle of where our planet got its water. I was surprised to find out just how knotty that last question is, and how our latest insights may have a bearing on where else in the galaxy life might be able to appear.

What follows, then, is the story of the best and brightest meteorite hunters, the treasures they have found and the secrets inside them. Put it all together, and you have a tale 4.6 billion years in the making.

JOSHUA HOWGEGO

MAY 2024

Among the hills a meteorite
Lies huge; and moss has overgrown,
And wind and rain with touches light
Made soft, the contours of the stone.

Thus easily can Earth digest
A cinder of sidereal fire,
And make her translunary guest
The native of an English shire.

Nor is it strange these wanderers
Find in her lap their fitting place,
For every particle that's hers
Came at the first from outer space.

From *The Meteorite* by C. S. Lewis

PROLOGUE

In the Basement

ONE SPRING DAY, I took a train and two buses to London's Natural History Museum. I had been many times before to see the dinosaur fossils and other delights, but I had previously always stuck to the public galleries. This time, I was going behind the scenes to where the scientists work and the bulk of the institution's treasures are kept. Martin Suttle, a researcher based at the museum at the time, had agreed to be my guide. We walked through a long gallery bustling with visitors to an exit. Suttle swiped us through with an electronic key card, closed the door and the hubbub died away.

We went down a flight of stairs to the basement, the air tinged with that scent of old paper and mildew you get in very old buildings, and along another long corridor lined with cases of impressive mineral crystals. And some way down, we reached the heavy wooden door to the museum's meteorite collection.

Inside, the space is modest and largely taken up with ranks of wooden drawers that rise from floor to ceiling.

I was in the early stages of researching meteorites at the time, and I had asked Suttle if he'd mind showing me around. He now gamely did so, opening drawers, pulling out samples and letting me weigh some of the extraterrestrial stones in my hands.

Each of the rocks in this room is a verified meteorite, which means it is among the oldest things you can possibly touch. They formed in the solar system's first flush of youth and eventually ended up crossing Earth's orbit at just the right moment. They then seared their way through our atmosphere, travelling at up to 650 kilometres per hour. Depending on their size and what minerals they contained, they may have etched fleeting green fireballs in the sky. Friction from the atmosphere heats the rock to incredible temperatures, often creating a blackened fusion crust on the outside. The heat can also break apart the stone in mid-air, so that fragments of it rain down over a long tract of ground. This area is known to meteorite hunters as the 'strewn field', a term I find highly delightful. Only once they have struck the ground do we call these stones meteorites. While soaring in the air, they are meteors (or 'shooting stars'). And when they are still in space, we call them meteoroids – or asteroids, if they are larger than a few metres wide.

As Suttle talked me through the treasures in the drawers, I kept hearing words like 'chondrites' and 'Imilac' drifting at me, and I wondered how often I could realistically interrupt to explain I wasn't entirely following. (I should add that I haven't yet discovered the limit of Suttle's patience with my questions.) Since then, I have gradually learned the special terminology used to talk about the plethora of meteorites in our collections. My aim is to spare you all but the most essential jargon in the pages that follow, but still, it will be helpful to prime ourselves with the basics.

Planetary scientists classify meteorites in a manner akin to how biologists classify life. Living things are divided and subdivided into kingdoms, orders, families and, ultimately, species, in a structure that resembles a family tree. It is similar with meteorites. But rather than genes and physiology, what separates one meteorite from another is their chemistry and the conditions under which they formed and changed over time. The study of these characteristics can be extremely revealing, as we will see.

At the top of the family tree sit two main groups: meteorites that have never melted (called chondrites) and those that have melted (called achondrites). As Suttle showed me in the museum basement, it isn't too tough for even an amateur like me to tell one from another – if the meteorite has been split open to expose its insides. Do this with a chondrite, and you can see that the rock is suffused with minute dots in whites, greens, browns, greys. These specks are called chondrules, and they are the first solids that formed in our solar system, in a cloud of dust encircling the young sun. Over time, gravity squished chondrules together to form pebbles, and then larger rocks, and, if those rocks never melted, then the chondrules are still visible inside meteorites today. The achondrites don't have chondrules inside them any longer because at some stage they have melted, perhaps because they were once part of the sizzling core of some huge and long since destroyed asteroid.

Some of the subdivisions of these two main groups are handy to know about too. The chondrites are, among other things, divided into 'ordinary' and 'carbonaceous' chondrites. The former are rocky and considered slightly boring, even among meteoriticists, because they are extremely common – the blue tit of meteorites. The latter are made of black, charcoal-like material that is packed with water and carbon-based molecules, the raw materials of living things. We will come across these rare birds many times in the book because they hold clues about how life got started in our solar system.

The achondrites are split into three groups: stony, irons and stony-irons. The 'stony' types are exactly what they sound like. The iron types are almost exactly what they sound like, except iron from space isn't quite the same as iron on Earth. In the basement room, Suttle showed me a slice of an iron meteorite with beautiful, silvery zigzagging lines running through it. These are known as Widmanstätten patterns, and they are the result of the way the iron crystallised as it cooled in space. The crystals in iron meteorites can be several centimetres long, a state they reach only if they cool exceptionally slowly, over millions of years. This never happens on Earth, meaning that these zigzags are a sure sign that the iron is extraterrestrial. The fact that some meteorites contain iron and some don't explains why metal detectors are a popular tool for some of the meteorite hunters we will meet in this book – but also why they aren't a fail-safe way to find every specimen.

The stony-irons are my personal favourite variety of meteorite. The thought is that these might have originated from a burgeoning planet, which had grown large enough so that its own gravity would produce an internal structure similar to that of Earth's today, with a heavy metallic core surrounded by a lighter rocky layer. If that planet was smashed apart by another asteroid, that would free fragments from the boundary between the rock and the metal, producing a stony-iron meteoroid. Whatever their origins, these meteorites are often beautiful, featuring transparent minerals mixed with the metal in an irregular honeycomb. I noticed one slice of such a meteorite in a corner of the basement. It turned out to be a piece of the Imilac meteorite, found in Chile in 1822, which had been mounted and lit from behind, so that the metal glinted as the light streamed through the greenish minerals. I thought it looked like a kind of alien stained-glass window.

To understand where meteorites come from, you must go back about 4.6 billion years, to the solar system in its infancy. The sun had recently begun to shine and surrounding it was only a wispy disc of dust and gas. Gradually, under the influence of gravity's attractive force, that dust and gas began to form tiny grains, then specks, then stones, then boulders that would eventually coalesce into planets. But some never quite made it. Today, we call those failed planets asteroids. Many are scattered in orbits between Mars and Jupiter, the fourth and fifth planets from the sun. It's not like that scene from *Star Wars*, where Harrison Ford pirouettes the *Millennium Falcon* through a dense field of jostling space boulders. The asteroid belt is sparsely populated and even if you combined all the rocks they would still weigh less than the dwarf planet Pluto.*

Sometimes, asteroids would bash into one another and pieces would splinter off. In this way, the solar system became filled with modest-sized space rocks that would orbit the sun, sometimes in neat circles and sometimes in wild ellipses, sweeping out into the cold outer reaches of the solar system and then dipping back inwards. Only the tiniest fraction of these stones would ever chance to meet planet Earth, but over the long life of the solar system some inevitably would. When these stones approach our skies, they have been cold as cold can be for billions of years. Then, suddenly, they hit our atmosphere and become meteors.

Some incredibly rare meteorites have a different origin story and belong to a wholly different branch of the family tree: the so-called planetary meteorites. These come not from asteroids, but the moon and Mars. In principle, they could come from other rocky planets or moons, such as Venus, but, if they are out there, we have not yet found – or perhaps recognised – them.

---

* What is the definition of a 'proper planet', I hear you ask. It is a question that has divided astronomers, sometimes bitterly. When I was a child, I learned that the solar system had nine planets. I even learned a mnemonic to remember them all: 'My Very Easy Method Just Speeds Up Naming Planets' (for Mercury, Venus, Earth, Mars, Jupiter, Saturn, Uranus, Neptune and Pluto). But these days Pluto is no longer officially classed as a planet. Mike Brown at Caltech discovered that there are quite a few other bodies beyond Neptune that are similar in size to Pluto, including a body called Eris. Astronomers had a choice: did we want to expand significantly the number of planets in the solar system? Or did we need to have a long talk about what really counts as a planet? In the end, a new definition was agreed upon and Pluto did not make the cut. To this day, Brown uses the handle @PlutoKiller on social media.

At the bottom of the family tree, meteorites are split into the finest subdivisions. Here the groupings are made on the basis of chemistry. No yawning at the back! This stuff matters because when rocks have very similar chemistry we can deduce that they must ultimately come from the same asteroid (or at least a family of asteroids that were once all part of the same parent rock). These groups go by obscure-sounding names like the 'L-chondrites' for example, or the 'CM-chondrites'. We tend to require at least a few chemically similar specimens before creating a new group, which means there are also many ungrouped meteorites, which – for now, at least – remain lonely, with no siblings and no place on the family tree.

Finally, individual meteorites are named after the place where they are found. This always used to catch me out, because meteorites can fall in remote places you've never heard of. Take the Nedagolla meteorite, which we will come across in the second half of the book. When I first read about 'Nedagolla', I half-guessed it was some obscure grouping of meteorites I'd not encountered before. In fact, it is the name of a place in India.

As I stood there with Suttle, I reflected that all the hundreds of meteorites stored in this repository had, at some point, been found. Sometimes, yes, by chance encounter, but more often by explorers and scientists determinedly on the hunt. I was itching to find out more about who they were and how they operate. But before that, we need to know how people first understood the bald fact that stones from space can sometimes fall to Earth. After all, for those who lived before scientists or quirky museum displays could prove otherwise, it was an extremely outlandish proposition.

PART ONE

The Hunt

CHAPTER [illegible]

The First [illegible]

The Hunt

It was [illegible] December [illegible]
and [illegible]
the fields [illegible]
sounds like explosions from [illegible]
he looked [illegible] saw something [illegible]
[illegible]
[illegible]
[illegible]
[illegible]
all around [illegible]

We know very little about [illegible]
[illegible]
[illegible]
[illegible]
[illegible]
[illegible]

[illegible]
[illegible] I wanted to walk [illegible]
Newton, the village where [illegible]

CHAPTER ONE

The First Meteorites

IT WAS A mild December afternoon in 1795 in Yorkshire and a ploughman named John Shipley was out working in the fields as usual. At about 3 p.m., he suddenly heard loud sounds like explosions from a gun above him and then, as he looked, he saw something falling from the sky emitting sparks of fire. Before he could make a move, this thing, whatever it was, smacked into the ground within spitting distance from where he stood, throwing up a shower of earth and small stones that pitter-pattered to the ground all around.

We know very little about Shipley, apart from this smattering of facts from one afternoon in his life. But this bizarre event made sure that his name would endure. It turned out to be a pivotal moment in the twisting story of how we came to accept what had long seemed completely unbelievable: that stones from space can sometimes just fall from the sky.

How would the ploughman have felt that fateful afternoon? I wanted to walk in his footsteps, so I drove to Wold Newton, the village where all this happened. There's not much of it, just a few houses, an eleventh-century church and a large duck pond. On the outskirts is Wold Cottage, which, despite the name, is a grand house with rolling farm land around it that Shipley once tended. The property is now a plush bed and breakfast, and as I drove towards it I saw neatly trimmed lawns and hedges, beds stuffed with flowers and a geodesic dome that serves as a greenhouse. I had asked the owners' permission to visit the spot where the stone fell. The weather was awful: freezing cold with a stiff wind and steady rain. Still, I had come this far. So I pulled on a woolly hat and a windcheater, got out of my car and trudged into the countryside.

I walked along a rough track that runs beside an upward-sloping field strewn with haystacks. Up in the corner was a tall, slim tower of red brick, a monument to the Wold Cottage meteorite, as the rock that nearly killed Shipley has become known. There was an incredible sense of openness, fields stretching out in all directions, almost no sign of humanity. Wold Cottage itself was hidden behind a hedge and I could only see a few tiny buildings on the horizon. I thought this scene couldn't be much altered compared with Shipley's day.

As I stood there in the wind and the rain, I tried to put myself in the man's shoes. I thought about how the chances of a meteorite hitting any given spot on Earth are minute and mused on how improbable Shipley's close encounter was. I'd wager he was terrified and probably in shock. Reports from the time say that he thought heaven and Earth were somehow coming together. Once he recovered his wits, did he question what he thought he'd just seen? Had a stone really fallen from the sky – and, if so, where in all the world had it come from?

We now know that this stone was a meteorite and that they can and do fall to Earth. But hundreds of years ago it was considered such a crazy idea that no one gave it the time of day. Shipley, who had narrowly avoided being killed by a falling space rock, was a critical witness to the contrary. What is more, his close call happened at just the right moment. At the close of the eighteenth century, a handful of strange episodes involving 'falling stones' took place all over Europe, finally forcing people to properly investigate what was going on.

Having said all that, there is evidence that people living many centuries ago knew about stones falling from the sky. To see that, we need only look at one of ancient Egypt's most celebrated pharaohs, Tutankhamun. When his sarcophagus was opened in 1925 there was a long, thin dagger on his right thigh. The blade is made of iron and the sheath is gold. But there is something distinctly odd about finding an iron dagger in a 3,300-year-old coffin. Iron wasn't smelted out of ore in Egypt until some 300 years later. That dagger should not exist.

In 2016 scientists looked at the blade and discovered another strange thing about the dagger – although it is mostly iron, it contains nearly 11 per cent nickel. This is odd because nickel is rare on Earth's surface. We also know that around the time of Tutankhamun's rule a new hieroglyphic phrase came into use, which translates roughly as 'iron from the sky'. It seems that the boy pharaoh's ornamental weapon was fashioned from a meteorite.

There's evidence that other ancient societies also knew about meteorites. Take the nomadic peoples who lived on the Gran Chaco plain of central South America at the time Europeans first visited in the sixteenth century. These people knew of a huge lump of iron lying in the middle of the wilderness, now known as the Mesón de Fierro (which translates as 'Large Table of Iron') in present-day Argentina. They told the Spanish that the stone had fallen from the sky and their name for the region translates as 'Field of the Sky'. We now know that the Mesón de Fierro is indeed one part of a massive meteor that fell to Earth about 4,000 years ago.

Back then, people understood and accepted that stones fall from the sky. But as the world became more connected over the centuries, something was lost. The Spanish completely rejected the idea that the Mesón de Fierro could have come from above, perhaps out of a hesitancy to believe something as outrageous as this based on the word of supposedly uneducated people. But then things began to shift again. In the mid-sixteenth century, most Europeans still believed that the Earth was the centre of the universe. But in 1543, Nicolaus Copernicus published *On the Revolutions of the Celestial Spheres*, his epoch-defining work that argued that the Earth orbited the sun. And it wasn't long before the Enlightenment got into full swing. People realised that we could collect evidence by observing the world – and the heavens – and thus inch our way to a truer, more scientific grasp of how reality works. Scientists have been doing as much ever since, and it has given us a far richer perspective on our place in the cosmos. We now know our sun is one of an estimated one septillion others in a universe that is at least 94 billion light years wide. The evidence of meteorites continues this tradition, as we will see.

Thanks to this adjustment of attitudes, people in Europe didn't want to blithely accept the wisdom of the ancient Greeks any more; they – rightly – wanted to believe only what could be empirically proven. That meant that when there were reports of stones falling from the sky from time to time during the seventeenth and early eighteenth centuries, almost no one was willing to accept them at face value.

Consider what happened near the French town of Lucé on 13 September 1768. According to contemporary reports, several harvesters were startled by a sound like a clap of thunder, followed by a loud hissing noise. They looked up to see a stone plummet to Earth. The French Royal Academy of Sciences heard about this and asked a trio of its chemists to investigate. One was a young Antoine Lavoisier, a man who would go on to make crucial discoveries that laid the groundwork for modern chemistry. Lavoisier and his colleagues conducted a chemical analysis of the rock. They concluded that it was an ordinary terrestrial stone that had been struck by lightning, which explained its charred, blackened surface. There was no shortage of reports of this kind. But testimonies that rocks had fallen from the sky – especially from poor labourers – were invariably dismissed. The scientific community did sometimes investigate, but it alighted upon plausible-sounding explanations that didn't involve rocks falling from space.

This all began to change in 1772, when Peter Simon Pallas, a German natural historian, set out on his travels in Siberia.

There, in a remote village around a hundred miles south of the city of Krasnoyarsk, he came across a huge and very strange-looking mass of iron that had been found by a local blacksmith some twenty years earlier. This thing really did look weird. Pallas described it as being 'as rough as a sea sponge' and riddled with cavities, many of which were filled with a transparent, amber-coloured material. He doubted the villagers' tales about this stone being a gift from heaven, but he did two things that ensured that many people heard about it. First, he had the stone transported to St Petersburg and displayed in the Kunstkamera, a hall of curiosities. Second, he wrote about the bizarre stone in a widely read book on his travels in Siberia. Learned people in Europe slowly began to talk about this exotic stone and wonder what it was. Today, there is a whole class of meteorites called pallasites after this specimen and they are beautiful to behold. They are often displayed with a backlight, so the light shines through the minerals encased in the iron, like the meteorite I saw in the basement of London's Natural History Museum.

The man who would do most to make sense of Pallas's find is an unlikely character. Ernst Chladni was born in 1756, the son of a law professor at the University of Wittenberg in Germany. The best-known engraving of him shows him smart and solid-looking, with short wispy hair and a double-breasted jacket. He earned a doctorate in law at the age of twenty-six but this was mainly at the insistence of his father, who died promptly after he graduated, leaving him free to pursue what had really interested him all along: physics.

Chladni was particularly interested in acoustics, the study of how sound moves through materials. No one had yet troubled to investigate this properly and Chladni soon made some delightful discoveries. One of his enduring achievements was to invent a technique to illustrate how solid surfaces vibrate in resonant patterns. This is much more fun than it might sound. Chladni would take a smooth metal plate of a particular shape – a triangle, a square, a pentagon – and scatter sand lightly over its surface. Then he would take a bow, like the one used to play a violin, in order to 'play' the plate and set it vibrating. After a few strokes it would reach a resonant note and – snap – the tumbling grains of sand would assume a pattern according to the sound waves that were coursing through the plate. These patterns were surprisingly intricate and beautiful and were always the same for a given size and shape of plate. Chladni wasn't the first to discover these nodal patterns, but his technique for demonstrating them made waves.

Chladni had always wanted to travel and, with his father gone, he needed money. So he came up with the idea of touring Europe and giving demonstrations of his vibrating plates, acoustics experiments and the few musical instruments he had also invented. He had his own coach made, with enough space to carry all his equipment. The lectures were successful – years later he would even demonstrate his plates in front of Napoleon Bonaparte. Yet Chladni's life was also tinged with sadness. He never received a coveted job at a university, and was forced to tour his acoustic show well into old age to earn a living. In a letter dated 1824, the scientist Wilhelm Olbers, based in Bremen, Germany, wrote: 'Dr Chladni is here again to give lectures on acoustics [...] I have gladly given him his fee; only with the understanding that he will not require me to attend every one of his twelve or fourteen lectures.' Chladni was sixty-seven years old at the time.

One day in 1793, Chladni was in the German city of Göttingen and had a conversation with Georg Christoph Lichtenberg, a prominent scientist, or 'natural philosopher' as they were then known. Lichtenberg had witnessed a large fireball a few years earlier and he would also have heard about Pallas's iron and the stories that it had come from the sky. According to a letter Chladni wrote years later, Lichtenberg told him that 'if all circumstances about fireballs were considered, they could best be thought of not as atmospheric but as cosmic phenomena'. This was one of the first serious conversations in which the idea that meteorites come from space was aired. As a result of this chat, Chladni went straight to the Göttingen library, where he spent three weeks finding out everything he could about atmospheric fireballs and falling stones.

He identified about twenty-four accounts of fireballs that had been observed between the late seventeenth and late eighteenth century, one of which was the Pallas 'ironmass', as it was then known. The accounts spanned more than 100 years and multiple continents – and what struck Chladni was how similar they all were. There were descriptions of explosions, and the sizes of the rocks were all quite similar and so were their speeds and flight paths, as far as he could tell. Steeped in legal training, Chladni felt it was unlikely that all these people would have made up stories so alike in the details. He thought the eyewitnesses were telling the truth. He set this all out in a book published in German in 1794; it has a very long title so it tends to be known simply as *Ironmasses*. In the book, he also argued that the speed of the falling stones ruled out a terrestrial origin. These stones, he contended, must have come from space.

It took a long time for ideas to spread in the eighteenth century and Chladni's book didn't receive much attention straight away, especially outside the German-speaking world. But over the next few years, three remarkable meteorite falls would force Chladni's hypothesis into the limelight.

Two months after the book was published, a strange sequence of events took place in Italy. Late at night on 15 June 1794, the lower flank of Mount Vesuvius abruptly split open. Massive explosions removed the top cone of the volcano and a plume of black smoke and ash rose high into the air. Huge volumes of lava spilled from vents in the volcano's lower slopes and seeped into the outskirts of Naples. It was an eruption of rare violence, even for such a mighty volcano. As the commotion continued the following day, something even more singular was about to happen 450 kilometres to the north in the town of Siena.

At this time, Siena had a population of around 30,000, and in June it was swollen further by an influx of rich English tourists. In the early evening, many of them reported the appearance of an intensely dark, high-altitude cloud emitting smoke, bright flashes and lightning bolts. Then there was a tremendous explosion and a volley of stones hurtled down from the sky.

Lots of people saw this happen. One stone reportedly shot through the rim of a boy's felt hat, leaving a scorched hole. Two English ladies saw stones land in a pond, splashing out puddles of water that appeared to be boiling hot. Some of these details could be embellishments, but this was an account of falling stones that was harder to doubt than usual, since so many people had seen it happen. Ambrogio Soldani, a monk and professor of maths at the university in Siena, collected many eyewitness accounts of the falls and examined nineteen of the stones, the largest of which weighed just over three kilograms.

Word of the Siena stones began to get out to people who could make a noise about it. Like William Hamilton, a man chiefly remembered today as the one-time husband of Emma Hamilton, a famous model who went on to be the mistress of Horatio Nelson. William Hamilton had a mind for the sciences. He was the English ambassador to the Court of Naples and while living there he made diligent observations of Vesuvius, in the process helping to lay the foundations of volcanology as a science. People had already started to ask whether the rain of stones in Siena and the eruption in Naples were a coincidence or if the latter could have somehow caused the former. An English expatriate living in Siena sent Hamilton one of the fallen stones, together with an account of what had happened, and this set his mind whirring.

One idea was that stones ejected by Vesuvius could have flown through the air all the way to Siena. Hamilton seems to have doubted this from the get-go, but he did venture up Vesuvius to look for stones that resembled those that fell in Siena. He didn't see any that fitted the bill, though admittedly the ground was covered in a layer of fresh ash. Writing in a journal of the Royal Society, Britain's foremost scientific institution, he floated another idea: might the ash from the volcano have drifted northwards to Siena and then been zapped by the electrical storm in such a way that it turned to stone and dropped to the ground? The fact that the stones were clearly blackened on the outside seemed to lend weight to this idea. The incident of the falling stones in Siena had sparked widespread debate on meteorites for the first time. But before anyone could get to the bottom of the mystery, something strange happened in Yorkshire.

It was a matter of months later, in December 1795, that John Shipley got the shock of his life while out in the rolling fields near Wold Newton. After the rock had fallen and he had come to his senses, he and two other men, a groom and a carpenter, went to investigate. They found the stone had seared through nearly half a metre of soil and another 15cm of solid chalk underneath that. They excavated the smelly hunk of rock and found it was a whacking great thing, the best part of a metre wide.

It is possible that Shipley's story could just have been told in the nearby area and would never have gained much credence further afield. But Shipley was employed by a man named Edward Topham, and that changed everything. Topham wasn't just wealthy, he was well known. In the 1990s, the space scientists Colin and Judith Pillinger at the Open University, UK, took an interest in Topham's story. Based on their historical research, they sketched a picture of him as a man who led revolts at school over the privileges of the prefects, who supported his friends in duels over points of honour and who defended the newspaper he owned in later life against accusations of defamation – and won. Topham was a man who stood up for what was right. It seems he spoke loudly, and when he did, people listened.

Topham wasn't at home when the stone fell, but on returning to Wold Cottage he collected statements from several people who had seen or heard the stone fall, including Shipley, and sent these off, together with his own reported description of what had happened, to a number of magazines and newspapers. He also exhibited the stone in Piccadilly, in the heart of London, across the street from a fashionable coffee house. Many people came to gawp at it, including Joseph Banks, a famous botanist and the president of the Royal Society.

Following the Wold Cottage incident, people really got talking about falling stones. In 1796, a fellow of the Royal Society named Edward King published the first English language work, a short book, about falling stones. In the same year, the natural philosopher Marc-Auguste Pictet founded a journal of the sciences and arts called *Bibliothèque Britannique*, and he began to publish many articles discussing meteorites and the swirling theories about where they came from. One popular idea was that the stones somehow accumulated in the atmosphere. Around the same time, translations of Chladni's book arrived in western Europe and, according to the late meteoriticist and historian Ursula Marvin, 'the trickle of journal articles about meteorites rose to a flood'.

Pictet himself wrote a piece saying how astonishing it was that many of the stones appeared so similar, with blackened crusts and glinting granite-like interiors. On the other hand, the object Pallas found in Siberia attracted a lot of discussion from some writers for being distinctly not like that at all: it was a huge lump of iron. Some would buy the idea that rocks could form in the atmosphere, perhaps triggered by electrical storms – but it was harder to swallow the same explanation for a giant lump of metal. And the fact that the Wold Cottage fall had happened on a day free from storms further detracted from that idea. One writer in particular, Guillaume de Luc, was having none of it. He insisted that the Wold Cottage meteorite never fell at all – were we really to accept the word of a labourer? – and that the Pallas iron simply had to be either from a volcano or the result of some ancient and long-forgotten mining and smelting operation. In short, debate was raging and there was little clarity.

In 1798, Banks received a letter about another fall in Benares in modern-day Uttar Pradesh in India. His correspondent told him that a shower of stones had fallen and that these all had blackened crusts and an internal structure composed of shiny grains, exactly like the Siena and Wold Cottage meteorites. This letter was a trigger for Banks. Enough was enough, he thought: we need a serious investigation of these stones. So he commissioned an accomplished young chemist called Edward Howard to get on to it.

Howard acquired four samples of meteorites: Siena, Wold Cottage, Benares and a much older stone called Tabor that had fallen in 1753. These were all of the black-crusted, stony variety. Howard finally read out the results of his investigations at a meeting of the Royal Society in 1802. It was a well-attended event, because his reading was interspersed with exciting reports on Ceres, a newly discovered dwarf planet about a quarter the size of the moon that orbits in the jumble of asteroids beyond Mars. The clinching evidence that Howard had gathered was that all the fallen stones he analysed contained a significant amount of nickel, an element that – as we know from the analysis of Tutankhamun's dagger – is rare on Earth. Howard's results were reproduced by other scientists, and this did much to cut through the noise of the previous few years and convince people in Europe that fallen stones were indeed a separate type of matter that must originate beyond the confines of this planet.

In April 1803 there was a very special meteorite fall, this time in L'Aigle, Normandy. The sheer number of stones that fell was astonishing – a fireball shot through the sky and after three violent explosions showered nearly 3,000 meteorites into the fields below. It was also crucial because, at length, a French minister sent out a man named Jean-Baptiste Biot to investigate and he produced an especially careful report, detailing the area over which the stones had fallen. This enabled him to predict unsearched areas where stones could be lying and then go out and find them. It was masterfully done, and in July 1803 Biot read his report to the National Institute of Sciences and Arts, the name of the French Academy of Sciences at the time. Here, according to Ursula Marvin, it 'was taken as definitive proof that stones fall from the sky'. Scientists still could not agree exactly where in space meteorites came from. That would take many more years to figure out. But only a slim minority now clung to the notion that they were imagined by ignorant peasants.

Most of the Wold Cottage meteorite is now housed in a pride-of-place gallery in the Natural History Museum in London, alongside treasures such as a first edition of Charles Darwin's *On the Origin of Species*. But if you want to see a selection of the first meteorites in modern history, there is no better place to go than Vienna. The city's Natural History Museum has one of the best collections of meteorites in the world. In the summer of 2021, I published a magazine story about meteorites and soon afterwards the museum's curator of meteorites, Ludovic Ferrière, wrote to me saying he liked the article and, if I was interested in meteorites, maybe we should talk. I eventually found myself with an invitation to a behind-the-scenes tour of the museum in Vienna.

I took the subway to Stephansplatz, the square smack in the middle of the city and emerged above ground next to St Stephan's cathedral, the church where Mozart married Constanze. I decided to walk the rest of the way across town to the museums quarter. Horse-drawn carts clip-clopped the streets, while baroque architecture, fountains and sculptures gilded the pavements.

Ferrière greeted me at a side entrance to the museum and showed me to his office. It was exactly as you would imagine a meteorite curator's lair to be, the rooms high-ceilinged and reassuringly musty. We passed stacks of boxes of rock samples and desks strewn with papers, lamps and microscopes. There were dark wooden cupboards of different sizes everywhere and Ferrière showed me inside several of them; one contained row upon row of thin slices of meteorites, while another was stuffed with what he said was the oldest collection of mineralogy books in the world. They are exquisitely bound in an array of red, orange, green and brown leathers and embossed with gold lettering.

At the end of the office was a heavy wooden door that opened directly onto the noisy public galleries. Ferrière ushered me to the other side and then through several halls lined with cases of precious minerals, until we reached the meteorite room. 'Here you are in my kingdom,' he said, spreading out his hands.

It's quite a place. There are 1,100 meteorites on display here, making it the world's largest exhibition of its sort. It's also the oldest: founded in 1748 by Francis Stephen I, husband to the indomitable Holy Roman Empress, Maria Theresa. Two of the founding pieces in the meteorite collection are the Hraschina meteorite, which fell in 1751, and the Tabor meteorite, which fell in 1753 (the same one that the young chemist Edward Howard would include in his landmark analysis decades later). Both rocks were collected at a time when falling stones were considered curiosities with no clearly defined origin. The Hraschina meteorite in particular is incredible to behold – especially since Ferrière has installed it on a rotating turntable. It weighs nearly forty kilograms and is the size of a small suitcase. Jet black, it's incredibly smooth and covered in bubbles, like the inside of an Aero chocolate bar magnified many times.

In this room, you can see samples of the Siena, Wold Cottage and L'Aigle meteorites. The stones are set out in alphabetical order in waist-high display cases and I took my time to find each in turn. It would be easy to be underwhelmed by these stones; on casual examination they all look like small black rocks, and perhaps that goes a long way to explain why it took us so long to accept these really are extraterrestrial. Then I reminded myself that the blackening was caused by the stone burning its way through our atmosphere at a speed of several kilometres per second and all at once it did feel like a special moment, standing there in the middle of a room filled with over 1,000 stones that came from outer space.

And not every meteorite in the Vienna collection is small and black. The Hraschina meteorite, for instance, is clearly an iron. It makes me think what a happy accident it was that Howard's chemical analysis only looked at simple stony meteorites. If he had looked at the full diversity of meteorites that we now know fall to Earth, including the irons, perhaps he would have ended up muddying the waters even further. But that wasn't how history played out.

After the three crucial falls between 1795 and 1803, scientists were finally convinced that stones do sometimes fall from the sky. But it was a far longer and more difficult task to work out where exactly they came from. The idea that they came from volcanoes collapsed swiftly, since there were so many examples of meteorites falling nowhere near a volcano. For decades, though, scientists such as Chladni maintained that they must come from interstellar space, that is, beyond the confines of our solar system. We could measure the speed at which meteorites entered our atmosphere and it appeared to be so incredibly nippy that anything travelling that fast could not maintain an orbit around our sun – so it must have come from beyond. These measurements turned out to be mistaken, however, and today we know that meteorites come from within our sun's orbit. They come, in fact, from a range of places in the solar system, which is one reason why they are such incredibly valuable resources to draw on if we want to know how our cosmic backyard came to be the way it is.

It wasn't until around the 1960s that meteoritics became recognisable as the discipline it is today. At the time, many ambitious young scientists wanted to get their hands on the rocks they knew the Apollo missions would bring back from the moon. The way to demonstrate they were the right people for the job seemed obvious: they needed to become experts at handling and understanding other rocks from space, namely meteorites. And so, during that decade, scientists began to categorise meteorites in earnest and endeavour to tease out their secrets – but we will come on to more of that later.

After I had stood for long enough in that cold and rainy field in Wold Newton, I made my way back to the cottage itself, where I had booked myself a room for the night. Then I dumped my bag in a bedroom and spent the evening looking through the collection of old documents and paintings the owners had managed to save over the years. There was a wonderful old painting of Edward Topham, quill in hand, with a thick head of flyaway grey hair and unrealistically rosy cheeks. Another showed his three daughters in all their finery. But there is no picture of John Shipley, at least not that I have been able to discover.

The next morning I rose early, breakfasted and went into the village to visit the ancient church where I knew Shipley was buried. I nosed around in the sopping wet grass and found his gravestone. His name was clearly visible at the top, as was his age when he died – just fifty-one. The inscription was too weathered and green with moss to read and I didn't give it much thought, but I later found out the stone bears the following words: 'All you who do behold my stone, O; think how quickly I was gone, death does not always warning give, therefore be careful how you live.'

CHAPTER TWO

Modern Meteorite Hunters

SVEND BUHL GREW up on the outskirts of Germany's oldest city, Trier. At school, he learned that you could pick up Roman artefacts in the fields on the outskirts of town, so he often went out to search around, finding flint tools, bits of pottery, the odd Roman coin. 'There is something inside me that makes me want to find rare things,' he says. Some of his booty from those days is still stored at the city's archaeological museum.

Then came the day when he saw a meteorite for the first time in a glass case. Buhl no longer recalls exactly where or when that was, but he does remember what went through his mind. 'The thought struck me that they must come from somewhere,' he says, 'and I was always wondering where you could find them.' Buhl was in his early teens at the time, and it would be years before he found the answer to that question. But he would go on to become an acclaimed meteorite hunter and, through a series of perilous expeditions, amass his own impressive collection.

Buhl is far from being the only modern meteorite hunter. For some it is a passionate hobby and for others a full-time job. Each has their own tactics, preferred modes of operating and motivations – among which is the prospect of making an absolute fortune. Amid the throngs of space rock chasers, Buhl stands out as a responsible and thoughtful operator, but it can be a dangerous, cut-throat game. It is also becoming more and more popular, as information about meteorites becomes increasingly available. Some think we are reaching a tipping point where we desperately need new laws to hold back the throngs chasing these riches from the heavens.

Hunting for meteorites is, to a degree, like any specialist hobby, from stamp collecting to knitting. To the uninitiated, it can seem arcane and difficult. How do you remember all those stitches? What makes one stamp worth next to nothing and another worth thousands? With meteorite hunting, as with so many hobbies, you get started by doing your reading or simply getting stuck in, building up understanding as you go. One of the first things you learn about meteorites, for example, is that they often have a blackened fusion crust on their outsides, caused by their searing journey through the atmosphere. But that information on its own isn't much help when it comes to recognising a meteorite: there are lots of black rocks that haven't come from space and, equally, many meteorites have had their fusion crust beaten away by the elements. Meteorite hunters need a bit of tacit knowledge to help them get started. That said, there are tricks of the trade that will help. I once heard of a meteorite hunter who carried a walking stick with a magnet stuck on the bottom – a handy shortcut for finding magnetic space rocks.

Buhl has never been a full-time meteorite hunter. These days he works for a semiconductor company and before that he had a varied career, including in the mining industry, where he learned how to prospect for minerals and lay explosives. He got his start as a meteorite chaser in the early noughties, when he got talking to a friend's father, a man named Rolf Poppinga, who had worked for the oil and gas company Schlumberger in Libya and had also travelled widely in the North African deserts in his retirement. When the older man told him stories of meteorites found in those deserts, Buhl pricked up his ears. His curiosity about meteorites had only increased over the years and he was itching to hunt for them. 'I wanted to just see if I could find such a rock from outer space for myself,' he says. It wasn't just about the rocks, though, truth be told. He also ached for an adventure. Unlike stamp collecting and knitting, hunting meteorites can't generally be done at your kitchen table.

Poppinga helped him contact some Tuareg people living in the Ténéré desert, a region of the Sahara that spans parts of Niger and Chad. These people sometimes acted as guides for tourists. Through emails and phone calls, Buhl struck up a relationship with some of them and they agreed to be his fixers for a trip to the desert. Among them was an interesting character called Mohamed ag Aoutchiki Kriska. As a young man, Aoutchiki had studied biology in Paris. When he returned home, farmers in his locality were struck by a plague of locusts and he was able to intervene using chemicals that prevented the insects from mating, greatly reducing the size of the swarm and the damage it inflicted. This earned him the ominous-sounding nickname 'The Locust'. Both in the years before and after Buhl encountered him, Kriska was involved in a Tuareg resistance group that opposed the Niger authorities. But Buhl says that back at the time they worked together, things were quieter and he saw him doing good work, like setting up schools for Tuareg children who otherwise had no access to education. Pictures of him in Buhl's memoir, *For a Fistful of Rocks*, show him resplendent in indigo robes from head to toe, with scholarly glasses peeping out from under his litham, the Tuareg mouth veil.

Poppinga – still hungry for adventure – agreed to fund Buhl's meteorite hunting trip to Ténéré, on the condition that he could join in the fun. So in 2002, the duo, plus Buhl's friend Jost Hecker and Poppinga's former colleague Juergen Meineke, flew to the southernmost air strip in Algeria and then drove to meet Aoutchiki at an isolated outpost close to the border with Niger. The expedition was fraught with risk. As the group travelled deeper into the desert they passed the shells of abandoned vehicles. Some had burned tyres next to them, which Buhl took as the owners' attempt to attract rescuers with smoke signals. Miscalculate or run out of fuel out here, and sometimes there is no one to help.

Aoutchiki proved to be a brilliant guide. At one point during their travels, he found the remains of a camp made by a Neolithic hunting party. The winds had blown away the sand covering heaps of stone tools and petrified bones and they picked up two arrowheads made of green jasper. Aoutchiki also commanded the respect of people in the desert. Further along, the meteorite hunting party encountered a group of smugglers who ferried cigarettes and fuel from as far away as Chad into Algeria, where – if they could dodge the border patrol helicopters – they made a decent profit. This could have been a dangerous encounter for Buhl, but the smugglers all recognised 'The Locust' and rushed to shake his hand.

To prepare for the trip, Buhl had spent weeks in libraries photocopying academic tomes on meteorites and reading as much as he could. He and his team had also examined satellite images of the area in order to identify places where the surface of the Earth was as old as possible, meaning more time had passed for meteorite falls to accumulate there over the aeons. Buhl's calculations suggested that one stone weighing at least ten grams should hit every square mile of the desert every 1,000 years. Since the desert is very old, there should be plenty to find.

What you need is a place where the black stones will stand out against a plain background. Eventually, Buhl and the team found themselves on a plateau where the surface was bright white limestone gravel. They drove in zigzag patterns in order to cover as great an area as they could, all the while scanning the ground with binoculars. At one point, they spotted a blackened pebble the size of a chestnut and the local guides expected Buhl to pronounce whether this rock was a meteorite. There was just one problem: he hadn't seen one in the wild before and wasn't sure (you never can be *completely* sure, of course, until the stone is properly analysed). The stone was blackened, as meteorites often are, and dimpled all over. Buhl took these indentations to be regmaglypts – depressions formed in the rock as searing hot air whipped against a meteor's molten surface as it fell – so he felt it was probably a meteorite. Unfortunately, later analysis would show it was only a piece of sandstone with iron mixed in – a false alarm.

Buhl and his team then descended into a part of the desert called the Tafassasset, where around noon temperatures reached a blistering forty-six degrees Celsius. They searched for days, sometimes rigging up a canopy between their two vehicles to provide a shady respite from the energy-sapping sun. It was imperative not to step barefoot on the sand – doing so would immediately result in a burned foot. But as they continued to search, one of the Tuareg fixers, who went by the name of Souleymane, found a small stone that really did look the part. Buhl could see fine grains of metal embedded in a charcoal-coloured body. Later, when the rock was analysed back in Germany, he discovered that this was a fragment of the Tafassasset meteorite, other parts of which had already been discovered in the same area. It was a rare kind of meteorite called a primitive achondrite, with a chemistry that does not fit with any existing group on the family tree. Scientists are still grappling with how to classify it today and it remains ungrouped. It would be easy to feel a tinge of disappointment at this discovery, but it is frequently said that the most exciting thought a scientist can have is not so much 'eureka!', but 'hmmm, that's funny'. When we find a piece of evidence that doesn't fit our preconceptions, it can be a sign that we have stumbled upon a clue to some new piece of knowledge – it might take time to come to fruition, but it is deeply exciting, nonetheless.

That and one other specimen were the only genuine space rocks Buhl found on his first trip, but as he returned to different parts of the Sahara and other deserts over the next few years, he amassed a highly respectable collection of meteorites. Along the way he also met a bunch of like-minded enthusiasts and built his own elite team, which he now calls Meteorite Recon. It's diverse, including two Russians who studied at Lomonosov University, a German friend who is a lawyer by profession and a Swiss entrepreneur already in possession of a vast meteorite collection – who also happens to be a very good cook. That, says Buhl, is an indispensable skill out in the field.

One thing Buhl struggled to find in the Ténéré desert was an iron meteorite. These are rare at the best of times, at around 4 per cent of all known meteorites, but in the Ténéré only about 0.2 per cent of recovered meteorites are irons. Why so few? One theory is that irons have a higher density than rocky meteorites, so they may be sinking into soft sand more easily and ending up partially hidden. Another possibility is that nomadic people living in the area have been picking up the irons for centuries to trade or to smelt into tools. According to Buhl, the first cold-hammered iron arrow and harpoon points from the area date to around 2,400 BC, meaning people have had thousands of years to gradually pick the desert clean of iron meteorites.

We can't speak of desert meteorites without mentioning the most jaw-dropping of all. This tale begins more than a century ago in Mauritania, another North African nation to the west of where Buhl was hunting. In 1916 it was a French colony, and a French army captain named Gaston Ripert was stationed in the town of Chinguetti. Remote and ancient, Chinguetti has a rich history of Muslim scholarship. The town served as a gathering place for the Islamic pilgrimage to Mecca, and at its peak in the eighteenth century it had twelve mosques and five libraries. By Ripert's time, it was still beautiful and atmospheric, but diminished, with dunes nibbling at its edges. One day, the captain overheard a conversation between some camel drivers about what they called the 'Fer de Dieu' (or 'Iron of God').

It was enough to furrow the brow and, curious, Ripert asked the head man of the town, Idi Ahmed Ould Seïn, about the gossip. Ahmed at first denied any knowledge of it and only at length did he consent to take him to visit this mysterious object, and then only on the condition of secrecy. Ripert was forbidden to carry any means of measuring the geography of their destination – no compass, no map. The pair set out at dusk and travelled overnight by camel. At dawn, they arrived at what appeared to be a huge hill, with one edge polished by the aeolian sand to a metallic, mirror-like finish.

In later communications, Ripert described this thing as an enormous, isolated metallic mass, partly covered by dunes. It was gigantic: forty metres high and 100 metres long. The captain scrambled to the top of the hill, where he found a 4.5-kilogram lump of iron, which was much later confirmed to be a meteorite. He also encountered some exposed 'metallic needles' at the summit and tried to splinter a piece off one of them by hitting it with the meteorite – but he found it was oddly ductile, bending but not breaking. His guide was anxious to be off, so the pair soon retreated to Chinguetti.

Ripert gave his story and his souvenir rock to a colonial administrator called Henry Hubert, who passed them on to Alfred Lacroix, a mineralogist at the French Museum of Natural History. It was Lacroix who rubber-stamped the rock as a genuine meteorite and introduced the idea that the forty-metre-high iron mass could be one too. In 1924 he read a paper on the subject to the French Academy of Sciences, which described Ripert's nocturnal journey and fabulous sighting. Lacroix ended his paper by saying: 'if, in effect, the dimensions given by M. Ripert are exact, and there is no reason to doubt them, the metallic block constitutes by far the most enormous of known meteorites'. The largest meteorite we know of today is about three metres across – a tadpole compared to the legendary Fer de Dieu.

Is 'legendary' the operative word here though? It would be easy to dismiss it all as a tall tale, but there are reasons not to. For one, there is the meteorite Ripert brought back, a lump of silvery metal the size of an outstretched hand, inset with gleaming crystals. How did Ripert get hold of this, if not from a bona fide strewn field?

Maybe he purchased the meteorite surreptitiously and concocted the whole story. But if so, what was his motive? Based on everything I have read about him, Ripert would have gained little by inventing this absurdist yarn. Then there is the most telling argument, which concerns those 'needles' that Ripert described. It would be a strangely precise detail to invent, but according to a paper on Chinguetti by the historian of meteoritics Ursula Marvin, there is an explanation for them. You can sometimes find weathered iron meteorites in which areas of nickel-rich iron (a mineral called taenite) are left standing after the neighbouring regions of less hardy, nickel-poor minerals have been worn away. These taenite spikes, Marvin conveys, tend to be ductile just as Ripert described – and this fact was not known at the time of Ripert's alleged discovery. If he cooked up this detail, that would be a coincidence almost as big as his putative meteorite.

Sounds compelling? Well, hold your horses. This is one of those ideas where the evidence can build or demolish your optimism depending on which aspects you look at. Take, for instance, the fact that there is no known impact crater corresponding to the Fer de Dieu. Plus, meteorite hunters have searched in vain for this prize for 100 years. Among many other searchers, the French meteoriticist Théodore Monod led several expeditions in the 1930s, and even offered a 1,000 franc reward for anyone who led him to the meteorite. In the 1990s, two British meteorite experts, Sara Russell and Phil Bland, also went looking, this time with a film crew in tow from the UK's Channel 4. In the resulting documentary, called *The Meteorite that Vanished*, Russell and Bland discuss the existence or otherwise of the giant meteorite in idle moments as they travel, though they never seem entirely convinced that the thing is there to be found. At one point, they are mulling over whether such a vast meteor could have survived a passage through the atmosphere without breaking into smaller pieces. 'If anything of that size could survive, then iron would be it,' says Bland in the documentary. But his tone doesn't exactly scream confidence.

The pair's most promising adventure was to follow up a lead from a pilot who flew over the area around Chinguetti years earlier and claimed to have seen the iron hill. Bland

and Russell went to the coordinates he provided and trudged over the dunes, taking readings with a magnetometer. If there was a gigantic iron meteorite lurking beneath them, the readings should spike as they passed above it. But there was no spike. The dunes rolled on, seemingly forever, and the duo returned home disappointed.

Today, there is one man who remains defiant about the chances of finding the Fer de Dieu. Robert Warren is now retired, but he previously had a career with a major oil company, which in 2018 dispatched him to Mauritania. The company wanted to prospect in the country and Warren was there to act as a government liaison. He told me his employer took security 'incredibly seriously', so while in Mauritania he travelled with personal protection officers and lived behind bullet-proof walls, which frustrated him. He wanted to see a bit of the life and culture of his new home. 'I've lived in many countries, and I always try to find something particular to each place that interests me – and in Mauritania it was meteorites,' says Warren. 'I started going to the bazaars and street vendors and asking if anyone knew anything about them. Eventually, having looked at a hundred rotten rocks, I began to find some.' In this way, he taught himself how to identify meteorites. 'It was a great little hobby to keep me occupied while I was living behind those ballistic walls.'

One thing led to another, and it wasn't long before he came across the story of the Fer de Dieu online. He was immediately hooked. 'The idea that somebody found what is many orders of magnitude bigger than the biggest known meteorite – and it's kind of sitting there out in the desert – that was totally compelling to me,' he says. 'So I did what I suspect many others have also done, and spent all my evenings on Google Earth trying to find the bloody thing sticking out of the sand.'

That did not prove to be the route to success, but Warren has made genuine progress. Like others before him, he realised that the dunes creep languidly across the desert and so the Fer de Dieu could now be hidden beneath the sands. After all, Ripert had reported that the iron hill was almost concealed by sand, even in 1916. Working with his brother Stuart Warren, a physicist at Imperial College London, and his brother's student, Ekaterini Protopapa, he used satellite images taken over the years to measure how fast the dunes in the area travel. It turned out to be about one metre per year. He also reasoned that the dunes would have to be at least forty metres high to obscure the mass of iron. Not only that, but he also made his own expedition to Chinguetti, to test how far one can get by camel in a single night. ('This is an area with a strong wind and it's not very flat. You cannot canter on a camel here,' he told me.) The small team has recently published a paper that took all these constraints and used them to narrow down the possible location of Ripert's fabled meteorite to a tight area.

Warren reckons that there is an easy way to check if the meteorite is there. In 2004, the Mauritanian government carried out an aerial magnetometer survey of a large region of desert, including Warren's area of interest, taking readings about every 700 metres. Being iron, the giant meteorite should have left a giveaway signal in this magnetic data. The trouble is that when Warren put in a formal request for access to the survey data he was stonewalled, despite his good professional relations with the government. Warren couldn't help letting his mind wander into the territory of conspiracy theory: what if the Mauritanian government wanted the prize for itself and was deliberately holding back the information? Publishing his paper was an effort to force the issue and it worked. He managed to get access to the data, albeit not from the government but by wrangling it from a student with whom it was previously shared for other purposes. Warren then analysed the data, searching for magnetic anomalies. It turned out that the only significant anomaly he found was inside the area of dunes that his team had already highlighted in their paper, which is the most significant new evidence in this quest for years. Warren is now hoping to organise a return expedition to Chinguetti. He wants to do a finer-grained magnetometer survey of the area in order to really work out if this anomaly has the right dimensions to be the Fer de Dieu. If it has, at some point someone will have to drill down and see if they strike iron.

For my part, I still don't know what to think about this tantalising legend. I'm not immune to its romance and majesty and I definitely felt a twinge of electricity run down my spine when Warren told me about his 'anomaly'. This is the stuff of Indiana Jones films. But one does have to weigh the chances of ever finding the treasure against the costs of this now century-long hunt. I asked Warren at what point he would give up the quest – but he struggled to give me a direct answer, instead rattling off a list of several lines of enquiry he still wants to pursue. I warmed to Warren; he came across as a capable and confident man who has learned to handle himself across multiple cultures. I also think it's worth some of us asking the biggest questions and pursuing even the craziest ideas. How many world-changing scientific discoveries would have been missed if we hadn't? But I also think there's a good chance the Fer de Dieu will prove to be a white whale – for Warren, as well as for others before him.

I got the feeling that it is the careful background research and analysis that is part of the appeal of meteorite hunting for Robert Warren. This business is a heady mix of adventure and scholarship. It's treasure hunting for clever people. And that is, I think, how Svend Buhl feels about it too. He and his Meteorite Recon team have amassed sprawling libraries of old books on the history of meteorite science and prospection, many now out of print. They are also students of the Meteoritical Bulletin Database, the official record of space rocks maintained by the Meteoritical Society, a not-for-profit group that promotes research and education around its discipline. As he immersed himself in the literature, Buhl felt the call to go meteorite hunting in the Atacama Desert in Chile. The historical records showed him that, unlike the Ténéré desert, this was a place where – for unknown reasons – an awful lot of iron meteorites had been discovered. He assumed that if some irons had been found, then others had probably been missed. Chile had other attractions too. It was reasonably accessible, thanks to the Pan American Highway running through it, and it had none of the political instability that has regularly characterised parts of the Sahara. To plan their new expedition, Buhl and his team turned in particular to a work from 1975 called *The*

Handbook of Iron Meteorites by Vagn F. Buchwald. It runs to three volumes and contains more than 2,000 figures – and it told the team a lot about what they thought they needed for a successful Atacama meteorite hunt. 'We read this kind of thing because for us it is sexy,' says Buhl.

In 2017, he and his team flew to Calama, a city in the north of Chile that acts as a gateway to the Atacama Desert. Driving their all-terrain vehicles into the wild, they were expecting a glut of meteorites. The surface here is so old and dry that Buhl compares it to Mars. The red planet has no plate tectonics or rain, so the many meteorites that have fallen over the aeons sit where they land, pristine in perpetuity. 'When rovers go to Mars they find an iron meteorite every few metres,' says Buhl. But things were not going to be quite that easy for him here.

The Atacama Desert is a remote, vast and beautiful place. During the trip, Meteorite Recon put a drone into the air to film their drive through the wilderness. The footage shows the team's two battered pickup trucks sweeping across a vast expanse of flat, red desert. Buhl leans an elbow out of the window, his face wrapped in a scarf and dark goggles over his eyes. This area is around 4,000 metres above sea level, some of it much higher, which is one reason it is home to some of the world's most advanced telescopes – there is less atmosphere to diffract starlight. The lack of air has its disadvantages though. Buhl found that when he climbed a small hillock, he was immediately out of breath. Worse, the normally powerful pickup trucks they had hired were far less muscular here than expected, due to the lack of oxygen entering the engines. In the Sahara, the blackened meteorites stood out against the yellow sands, but here everything was coated in a layer of fine red dust composed mainly of iron and aluminium ores. 'We found nothing in the first week,' says Buhl.

Despairing, Buhl and his team thought of the Imilac meteorite. We have met this one before in the basement of London's Natural History Museum, where I saw a beautiful fragment of it. It is actually one of the most famous meteorites, not just because of its good looks but also because so many fragments have been found over the years since its original discovery in 1822. Thanks to all these finds, the strewn field of this fall has been well documented. Buhl searched along some of the known strewn field with a metal detector and found a few small fragments of the meteorite, which briefly raised the group's sprits. They and others had long wondered if there could be more significant chunks of Imilac at the far end of the strewn field, which remained unsearched. These would be far greater prizes. Buhl selected a likely-looking area in the mountains and the hunt commenced.

Nothing. They left Chile empty-handed that year. However, when they returned in 2019, they managed to stumble upon a new part of the strewn field, which, it turned out, they had missed by just 300 metres on the previous trip. Then, it was merely a matter of walking across the area systematically with metal detectors. When the detectors pinged, Buhl knew it was a meteorite – the desert was so untouched that it was never a nail or a coin. Each of the team would find two or three small pieces of meteorite a day and though the search was hard work, the feeling of finding a meteorite was unbeatable. 'It was very thrilling, to have these meteorites that fell hundreds or a thousand years ago and be the first one to hold it in your hand.'

Buhl revels in the details of his work and shies away from the spotlight. Although he has written books about his adventures, he rarely gives interviews, preferring the private satisfaction of gleaning small slices of new historical perspective to the glare of publicity. For all his adventures, though, his favourite meteorite is one he found much closer to home. In 2023, he heard of a meteorite fall at Elmshorn, a few kilometres from his home in Hamburg, and went out on a whim to look for a piece of it. He wasn't holding out much hope, but to his surprise he found a chunk weighing eight grams after a short search. It was the element of surprise that made it so memorable, he says; that and the fact that this rock was so fresh. Most of his treasures fell to Earth many lifetimes ago, but this time he was holding a building block of the solar system that had been in the cold obscurity of space for billions of years – and had fallen to Earth literally yesterday. As we will see in the following chapters, picking up freshly fallen meteorites is many times rarer than finding old and crusty ones. This was a moment of awe that only a lucky few get to savour.

While that freshly fallen rock is an outlier in Buhl's collection, some meteorite hunters thrive on the rapid response side of the business. One is Robert Ward, a private meteorite hunter and dealer who lives in Arizona, US. As a child, he saw a huge fireball in the sky and it got him hooked on meteorites, so he read everything he could and started hunting and trading them. 'I did dabble in real estate, when I was younger, which is the family business,' says Ward. 'But I found I was making more money dealing meteorites. And I didn't have to pay property taxes.'

At home, Ward has a plush room with glass display cases arranged around the edge, each with several shelves of beautifully cut and polished meteorite specimens. In a room off to the side, he keeps several backpacks at the ready, each filled with clothing and gear for a different environment: cold northern latitudes, deserts, and jungle conditions. When news of an important new meteorite fall reaches him, speed is everything. If he can be there first, he can net himself some serious cosmic treasure.

On the morning of 24 April 2019, Ward got wind of a freshly fallen stone that would turn out to be particularly desirable. The previous evening, a little past 9 p.m. local time, a fireball streaked across the skies above Costa Rica. Residents of the village of Aguas Zarcas reported hearing an ominous rumble and seeing an orange–green trail above them. Pieces of the meteorite rained down, one smashing through the corrugated roof of a kennel belonging to a German shepherd named – by complete coincidence – Rocky. When Ward got the news, he immediately knew it was important. Photos of the meteorite showed a rock as black as coal with an iridescent shimmer. To the trained eye, this was evidently a carbonaceous chondrite, one of those rare meteorites that are rich in water and complex carbon-based molecules. If the ordinary chondrites are the blue tit of meteorites, these are the osprey. It was reminiscent of the Murchison meteorite, which fell in Australia in 1969 and was brimming with a plethora of molecules that are central to life's processes. There were nucleobases and the sugar molecule ribose, which are structural components of DNA, plus a raft of amino acids, the building blocks of proteins – the materials from which living tissues are built. Some credit the Murchison meteorite with helping to kick off the field of astrobiology, which studies how and where life could start beyond Earth. Ward had all of this in the back of his mind as he grabbed one of his backpacks and hastily booked a flight to Costa Rica.

Ronald Pérez Huertas and his family lived next door to the owners of Rocky the German shepherd. Having seen the damaged kennel and its inhabitant – which was rattled but thankfully unscathed* – they decided to make an all-out search for the meteorite. The next day, they pooled their finds and covered a table with fragments of space rock. (Ultimately, all the retrieved samples of the meteorite would amount to a total of twenty-seven kilograms.) After snapping a photo, they searched online for a meteorite dealer who might be interested. The list of hits returned one Michael Farmer, another well-known US-based meteorite

* Rocky's close shave might make you wonder if anyone has ever been killed by a falling meteorite. There have certainly been some near misses. At 2 p.m. on 30 November 1954, a woman named Ann Hodges was taking a nap on her couch in Alabama when a meteorite crashed through her roof and hit her on the thigh, leaving a nasty bruise. When her husband got home at 6 p.m., she reportedly told him there had been 'a little excitement'. There are plenty of stories from antiquity of people being killed by falling stones, but it is hard to verify them. However, in 2020 three scientists dug out records from Turkey that seem to be the only proper evidence of a death by meteorite. These show that on 22 August 1888, fragments of a meteor hit two men in Iraq, paralysing one and killing the other.

dealer whom Ward had previously worked with, and who has since largely retired.*

Ward and Farmer are cut from a different cloth to Svend Buhl. Ward often wears a large Stetson and calls himself 'the Space Cowboy'. In a promotional video on his website, he can be seen charging around chasing down meteorites and – for reasons left unexplained – firing multiple rounds from a handgun. Towards the end of the video he says: 'I know meteorites, I know danger, and I know how to have a good time.' To be fair to Ward, though, he laughed about this video when I interviewed him and came across as personable and professional.

In 2010, both men were on a meteorite hunting trip to Oman, where, among other things, Farmer found three pieces of a lunar meteorite. They had studied the law of Oman and believed their work was legal, but then one day they hit a roadblock. Men with M16 rifles arrested them and seized the meteorites in their car. The pair were taken to a jail and, in Ward's telling, were subjected to a terrible ordeal. Ward says he was thrown into a 'hole in the ground' and taken out to be interrogated at intervals. He believes the authorities thought he was spying and did not believe his story that he was hunting meteorites. At one point he was taken into an interrogation room that he says was liberally decorated with blood stains. Eventually, a very muscular man came in – Ward describes him as 'shredded' – and asked him if he was scared. Ward most definitely was. Later, back

* Farmer did not answer my emails or phone calls. The information about him in this story is based on previously reported information and conversations with people who know him.

in his cell, Ward says he heard what he believed to be the screams of a person being brutally beaten. Eventually Ward and Farmer were given a short trial and sentenced to six months in jail, where conditions were tough. 'I was offered less to eat than my pet cat eats in a day,' Farmer later told *New Scientist* magazine. With help from the US embassy, the men got out after two months, by which time Farmer had lost eighteen kilograms.

Aguas Zarcas also had its share of aggro. More and more meteorite dealers were turning up and competing fiercely with each other to buy the remaining stones. Farmer later told reporters that at one point he 'came close to sinking a shovel into one guy's skull'. But before long, he and Ward had bought all the meteorites they could and so they headed back to Arizona, Farmer taking a few larger ones and Ward taking more smaller stones. They paid a lot less than they knew they would be able to sell them for, according to reporting in *Science* magazine from journalists Joshua Sokol and Andrea Solano Benavides. I asked Ward about this and whether he ever feels bad about buying at low prices. He says that the people he purchases from often do have an idea of what meteorites are worth from reading online, and if anything their asking prices tend to be unrealistically high. Plus, for him, this is a business. The money he makes on a successful hunting trip has to cover the cost of his flights and all the expeditions from which he comes back empty-handed. After buying the stones the family had found, he went for a walk in their goat paddock and found another piece of the meteorite. 'I got that one at a reduced price because I found it myself,' he says.

Before Farmer left, he bought Rocky's doghouse, complete with the hole in the roof. He also has a well-publicised habit of eating a little of every meteorite he finds, including several lunar meteorites. He previously told reporters that he did this just so he could say he'd eaten a piece of the moon. The Aguas Zarcas specimen was, he said, one of the most unpleasant space rocks he had ever tasted. I can't help thinking this was an ill-advised snack, given the wild cocktail of carbon-based molecules present in carbonaceous chondrites.

Back in the US, Farmer shared some of the meteorite with scientists and one decidedly curious result has emerged from their analysis. It has to do with a subtle idea in chemistry called chirality. Many molecules associated with life can come in two mirror image forms. Think of these as being akin to your left and right hands (the word 'chiral' is derived from the ancient Greek word for 'hand'). These are reflections of the same chemical structure, though the two forms can have very different physical properties.* Curiously, all the molecules of life on Earth have a distinct chirality: DNA, RNA and their constituents are all right-handed, whereas amino acids and their constituent proteins are all left-handed. We do not fully understand why this is so. Perhaps, there was a fork in the biochemical path at some

* Chirality is far from just an academic curiosity. The classic reminder of this is the drug thalidomide, which was marketed as a cure for morning sickness in pregnant women starting in the late 1950s. The drug comes in two chiral, mirror image forms, and no one had realised that these have different effects: while one is a medicine, the other causes damage to foetuses. An estimated 10,000 babies were affected by the drug before it was withdrawn. Of those who survived beyond a few months, many had profound disabilities – including being born without arms or legs.

distant point in the past and something forced living things to 'choose' to use this system. Or perhaps there was already some physical factor that meant there were more molecules of a certain handedness around when life began.

In 2020, Daniel Glavin and his colleagues at NASA's Goddard Space Flight Center in Maryland studied the amino acids in the Aguas Zarcas rock. Looking at an amino acid called isovaline, they found there was about 15 per cent more of the left-handed form than the right-handed. It is certainly not case closed on life's mirror mystery, but this result and others like it hint that perhaps the dominance of left-handed amino acids might have been there in space, from before the planets even formed.

Scientists tend to have mixed feelings about private meteorite hunters. They can't generally drop everything and hop on a plane when a meteorite falls – those lectures won't write themselves – so having private meteorite hunters out there means there are many more specimens in collections than there otherwise would have been. Meteorite hunters often want their finds to be officially recognised and named, partly because this gives them proper provenance and adds to their market value. This means submitting a request to the Meteoritical Society, which entails the owner handing over a sample of their rock to be made available for study at an academic institution. (In the case of a rock weighing up to 400 grams, the society asks for either 20 per cent of its mass or 20 grams, whichever is the greater.) Under this system, everyone benefits – in theory. 'These hunters have done a massive service to the meteoritics community by providing samples, some of which are unique and important,' says meteoriticist Sara Russell, who is now based at London's Natural History Museum. 'But it is a double-edged sword.'

We can see some of the downsides of private meteorite hunting in North Africa, where Buhl's adventures started. Take the Moroccan town of Erfoud, a two-day drive east of Marrakesh. On the road approaching the town, a series of meteorite shops have sprung up. A reporter for the *Financial Times* visited a few years ago and talked to local people. One man spoke of how he would ride his motorcycle into the desert, pick up as many likely-looking rocks as he could and then take them to one of the meteorite dealers. Most would be worthless – but occasionally he would get lucky and sell a meteorite for the equivalent of several months' living expenses. There are many others doing likewise. Meteorite dealing has become a vital part of the local economy.

One reason for this popularity is the headlines about famous finds that dangle the promise of great riches. Back in 2011, for example, nomadic people found a meteorite in the Sahara that turned out to be a piece of Mars.* Nicknamed 'Black Beauty', it changed hands several times and was sold at auction in 2016 for almost £44,000. This, says Ludovic Ferrière at Vienna's Natural History Museum, has inspired many more people to go to North Africa in search of extra-terrestrial riches. 'It is like a gold rush, really,' he says. 'The

* For a chunk of Mars to fall to Earth, it first had to be blasted off the red planet's surface so explosively that it escaped the pull of its gravity. It is thought that this happened occasionally in the distant past when asteroids smashed into Mars; if they were big enough, the impact threw debris into space, some of which eventually crossed Earth's orbit. (So in a sense, meteorites beget meteorites.) We will come back to Mars meteorites again in Chapter Five.

main motivation for these people is money.' And though local people do glean an income out of all this, there are dubious goings-on too. Many of the meteorites are swiftly shipped out of Morocco and into private collections, stripping the country of its cosmic heritage. 'Is it right that they take these beautiful and precious things away from where they were found? I think it is a problem,' says Russell. Plus, people have been known to get lost, run out of water and die in the desert. Hunting meteorites out there is dangerous, even if you know what you are doing.

Another reason meteorite hunting is more popular is the rise of camera networks that track meteors. Over the past few years, many of these networks have sprung up in Europe. Cameras are installed all over the continent on interested peoples' houses and they watch the sky for incoming fireballs. This data is then pooled by the network coordinators, often a mixture of amateur astronomers and university academics, and the footage can be used to predict where the meteorite will land. Often, the result is a predicted strewn field, the swathe of land over which fragments of the meteor will have fallen. (We will dive deeper into the story of how these networks first became established and why they are so useful in the next chapter.) They often share their information publicly, so, for those paying attention, they provide a treasure map that makes the game of meteorite hunting much easier.

On a Sunday evening in early February 2023, a group of astronomers announced they had spotted an asteroid called 2023 CX1 that was on a collision course with Earth. At little more than a metre across, it was not dangerous, but it is extremely rare for scientists to recognise an incoming space rock before it even enters our atmosphere, and this one was so large that there was no doubt it would drop meteorites. Brigitte Zanda, a meteoriticist at the French Natural History Museum, was at home in Paris when her phone pinged with the news. She manages the French camera network Vigie-Ciel (which translates as 'Sky Watch'), and her team quickly jumped into action. At first, their predictions suggested the meteorites would drop in the English Channel, but as more observations of the asteroid came in during the night, they refined their calculations and it became clear that the strewn field would be in Normandy.

By the next day, members of the Vigie-Ciel team were combing for the fallen stones and they found a few small pieces during the week. The owners of the land were generally happy for the meteorites to be given to the museum, but as the week wore on more private meteorite hunters arrived on the scene. Among them was Steve Arnold, a well-known meteorite hunter from the US who appeared alongside Geoffrey Notkin in a documentary television series called *Meteorite Men*, which premiered in 2009. Arnold has a brash style and many scientists told me that he rubs them up the wrong way. In the television series, for example, there is a 'ker-ching' noise when he and Notkin find a meteorite and dollar signs pop up on the screen. 'Meteorites are sometimes viewed as objects of monetary value and only monetary value,' says Zanda. 'In my view they are objects of scientific heritage.'

Arnold and his fellow meteorite hunter, Roberto Vargas, heard about the incoming meteor while watching the

Superbowl and decided to fly to Normandy. They were there by the end of the week and, whether through luck, experience or a mixture of both, Arnold quickly found the main mass of the meteorite – tipping the scales at 178 grams – close to the town of Saint-Pierre-le-Viger. While out searching in a field, he noticed a hole in the ground and, when he looked inside, a meteorite was there, a few inches down.

What the duo did next really ruffled feathers. They repaired to a Normandy beach and posted a live video to Facebook, showing off the stone and chanting 'USA, USA'. During the video, Arnold explained that the stone was 'a little bit damp', not just because it had rained but because 'we did wash it off'. This is perhaps the single worst thing you could do to a meteorite from a scientific perspective, because it contaminates the pristine space rock with terrestrial liquid and would inevitably skew any analysis of its chemistry. Arnold and Vargas then had the stone sent to the US by courier.

Zanda says she always tries to remain polite, neutral and professional when dealing with private meteorite hunters. 'I am not searching for a fight with these people. We have to deal with them,' she says. She called Arnold and asked him if he would share the coordinates of where he found the stone and if she could get access to make some key measurements. According to Zanda, Arnold refused on both counts. While Zanda was at pains to be respectful to Arnold and Vargas, she felt deeply disappointed that hunters did not choose to share any of the mass with French academics. For my part, I think hunters like Arnold know the value of meteorites very well, but perhaps they have forgotten that showing kindness and respect costs them nothing. (By the way, I contacted Steve Arnold to ask for his comments on what happened in Normandy, but he did not respond.)

Sometime later, Zanda was at home one evening when her phone pinged again. It was a man named Bil Bungay, a science fiction film producer in London who is also a meteorite hunter. He felt the meteorite was of huge scientific value and was offering to try and buy it from Arnold – if Zanda was up for displaying it at her museum. She very much was. The details of the negotiations that followed are not in the public domain, but Bungay did get the rock from Arnold and it has reportedly been insured for £250,000. He has given it to Zanda's museum on a long-term loan, where it is now on display to the public.

You might think that the law would be on the side of scientists, but, well, it's complicated. It is illegal in France to access private land without permission, but in practice that is no help unless the landowner wishes to press charges. France doesn't have any law that governs the ownership of meteorites specifically, and that's not a rare state of affairs: few nations have any legal rules relating directly to meteorites. There are outliers, such as Denmark, where any meteorites recovered are deemed the property of the state. Finders must hand them over, and they are given compensation. In other nations, meteorites are in practice covered by laws designed to deal with mineral resources or related commodities and are sometimes seen as the property of the person on whose land they fall.

Recently, Zanda and Ferrière have been chewing over the design of hypothetical new and better laws. One option would be to make it illegal to remove meteorites from a country, full stop. That is the situation in Namibia, says Ferrière. But it does not prevent it from happening; it just means the stones are smuggled out and never officially recognised. 'A lot of meteorites are passing directly from Africa into private collections in China,' he says. For this reason, it may not be useful for laws to be too punitive. Ferrière thinks a better option is what he calls a 'win, win, win' model, where the finder, the scientists and the owner of the land a meteorite falls on are each given a share of the spoils.

Zanda has sketched out the broad strokes of a law along these lines and has conversed with various ministries in France about proposing it in the country's parliament, yet progress has been slow. Meteorites are not exactly a high political priority compared with education and hospitals, but Zanda says she is beginning to get some traction. Showing the ministries the video of Arnold on the beach was a shock tactic that got them interested. And then Vigie-Ciel tracked and retrieved another meteorite in the autumn of 2023. That was two falls in the country in a little over nine months. Zanda says this underlined the need for regulation and there is now appetite among her governmental contacts to at least look at a new meteorite law. Until and unless we have that, hunters of these extraterrestrial prizes in France will be able to choose what they do with their bounty.

For those specimens that don't end up in museums, many are last seen in public during an auction. In February 2022, the auction house Christie's held a sale entitled: 'Deep

Impact: Martian, Lunar and other rare meteorites'. James Hyslop, one of the auctioneers, told me that Christie's has been holding similar meteorite auctions for years now and that they attract collectors from around the globe. The lots sold one by one; a pocket-sized piece of the moon went for $12,000 and a beautiful slice of the Imilac meteorite made $27,000. And then there was a more unusual lot, courtesy of one Michael Farmer. It was Rocky's doghouse, complete with an eighteen-centimetre hole blasted in the roof. When the gavel came down, the sale price was $44,000. It is hard to know if that makes it the most expensive kennel ever sold, but it must be in the running.

CHAPTER THREE

Tracking Fireballs

I PROMISE YOU: I don't spend my entire life looking at rocks in fusty glass cases. But there was something I wanted to see back at London's Natural History Museum. So to South Kensington I went once again, this time through the grand entrance hall, where an articulated skeleton of a blue whale hangs from the ceiling and fossilised dinosaurs stand guard. I went up a grand staircase, past a marble statue of Charles Darwin, through a gallery and eventually found a room called The Vault, where the museum houses some of its most valuable treasures.

This place was like the inner sanctum of a Swiss bank. I stepped over the threshold of a sliding metal door that must be ten centimetres thick and into the small, brightly lit room, where I was met with some incredible objects. In one case sits the Devonshire Emerald, a dazzling shade of green and as big as the palm of my hand, and then there is a rare pigeon's blood ruby from Myanmar and a diamond-encrusted snuff box commissioned by the Russian Tsar Alexander II.

The object I've come to see seems a mite disappointing by comparison. It is a small lump of grey-black rock that sits in a large glass jar on a piece of silver foil, illuminated with spotlights. But when I press my face almost up against the glass, I can catch a kind of yellowish iridescence on the smooth sides. I'm looking at a fragment of the Winchcombe meteorite, recovered from a Gloucestershire field in 2021, which is the most recent meteorite recovered in the UK and easily the most important. It is also worth tens of thousands of pounds at the very least – I suppose that explains why they keep it in The Vault.

This meteorite is fascinating for lots of reasons. One is that, unlike most other meteorites, it contains a generous quantity of carbon-based molecules and water, the kind of materials that are needed to get life started. But even more important is the way in which it was found. Most meteorites are chance finds, picked up after having lain around in the dirt and the rain and the cold for decades. All that exposure leaves this kind of meteorite, called a 'find', in not exactly pristine condition. But meteorites which are seen to fall and are then picked up quickly – which are called 'falls' – are in another league. If we get to them quickly enough, they contain a highly authentic record of the conditions in the early solar system, unsullied by Earth's destructive weathering. The Winchcombe meteorite is a 'fall'.

It's true that we occasionally get incoming meteors that none could miss. Take the Chelyabinsk meteor, a stone that bombed through the skies close to the eponymous Russian city in February 2013. The original boulder is estimated to have been seventeen to twenty metres wide as it entered the atmosphere, and it then exploded in mid-air, creating an airburst with a force equivalent to 500,000 tonnes of TNT. It was, quite simply, an enormous bang, and one that was widely recorded on dashcams and mobile phones. One building collapsed and windows were blown in far and wide, the flying glass causing untold injuries. The explosion and fireball were so bright that seventy onlookers were temporarily blinded and twenty reported 'sunburn' – though perhaps that should be 'meteor burn'. But this was the largest incoming meteorite in more than a century. Most meteorites are much smaller and spotting them is a challenge not to be sniffed at.

The best way to track and then locate a meteorite is to take photographs or a film of it as it streaks through the sky and then use computers to project its trajectory forwards precisely enough so that we know where the rock will land. That sounds like a big ask, and it is. But there is a revolution going on among meteorite trajectory trackers, with networks of observing cameras around the world coming online to watch the night sky. It is these camera networks that helped us find the Winchcombe meteorite.

But that's not all. Just as we can project the trajectory of a meteor forwards, to help us find where it will land, so we can also project it backwards, to help us see where it came from. The hope is that tracking the trajectory of incoming meteors both forwards and backwards could help us to not only find the things more readily but also understand the history of the solar system, as written in meteorites, like never before.

People have been trying to record meteor trajectories for decades, long before video cameras or smartphones became widespread. In 1936, scientists at Harvard University began the Harvard Meteor Programme, which used long-exposure film to capture the trails of meteors through the night sky. Back then, we didn't even know for sure whether meteorites came from within our solar system or from interstellar space. At first, the plan was to use these cameras to study the atmosphere by watching the trail of a meteor and how it changed as it came down through the various layers. But then scientists realised that they could also reconstruct the meteors' orbits. The network of cameras expanded and, in the 1950s, the American astronomer Lawrence Whipple managed to show that the meteors' orbits were consistent with them coming from *inside* the solar system. Whipple made many other important contributions to astronomy, including the discovery of six comets – in fact, he used to drive around in a car with a licence plate that read 'COMETS'.

Alas, Whipple and his colleagues did not find their holy grail, never managing to actually recover a rock after tracking its trajectory through the atmosphere. Although they had calculated a number of meteorites' orbits, they couldn't study the stones themselves. They were out there, somewhere – under a bush or in some random field – the wind and rain gradually erasing their precious chemical records of the early solar system. What we needed was for someone to track a meteor's trajectory, forecast where it was going to land – and then actually retrieve it.

The scene for this scientific breakthrough would be 1950s Czechoslovakia. The Second World War was a recent memory and the country was part of the Soviet Union's

Eastern Bloc. Between 1949 and 1954, Joseph Stalin ordered the country's rulers to carry out purges of people deemed enemies of the communist way of life. It was a brutal era of torture, show trials and executions but Zdeněk Ceplecha managed to escape the worst of it. He was a young man in his early twenties, studying astronomy at the Ondřejov Observatory, half an hour's drive from Prague. Ceplecha didn't come from a bourgeois family and he kept out of trouble, so he managed to get on with his work.

Like Whipple, Ceplecha was interested in tracking meteorites. He set up cameras at two observing stations – one at the Ondřejov Observatory and another about forty kilometres away at the town of Sedlec-Prčice – which covered about half of the sky and took all-night exposures. The bright path of a meteor would show up as a streak along the glass photographic plate. To calculate the orbit of a meteorite, Ceplecha needed to know the angle of this streak and the time it happened. For the timing, he initially planned to rely on witnesses to record when fireballs appeared, which would have given only a rough time. Later, he added what's known as a sidereal camera, which tracks the motion of the stars as the Earth turns and so provides an accurate time stamp. By comparing the tracking cameras and the sidereal camera he would be able to determine the path of any meteors and the time they appeared and calculate a flight path.

For almost a decade nothing happened. Then, one night in April 1959, Ceplecha was at home near the observatory watching his black-and-white TV when, suddenly, out of the window, he saw an incredibly bright light in the night sky. He immediately knew it was a meteorite. The first thing he did was reach for a light meter, the sort used to set a camera's exposure. He used this to measure the brightness of the TV and then made a mental note of how much brighter the meteor had appeared. A meteor's brightness can sometimes be related to its size and he was hoping to use his estimate to do some calculations later.

Astronomers use a scale called apparent magnitude to talk about how bright objects in the sky appear. It is a reverse logarithmic scale, meaning objects get many times brighter for each step downwards on the ladder. The brightest star in the sky, Sirius, is set as −1; the full moon is about −12 and the sun is about −26. Ceplecha put the fireball that blazed across the sky that night at −19. For a few seconds, it would have lit up the land beneath almost as if it were day.* This was a big one. He ran out of the door and over to the observatory.

When he arrived, he found that the photographic plates had been ridiculously over-exposed by the extreme glare of the fireball. It was hard to make anything out. But over the next day or so, Ceplecha managed to make copies of the film that showed the rock's trajectory more clearly. As well as extrapolating the trail backwards into space, it could be projected forwards too, to show where the rock ought to have fallen. It had been heading towards the town of Pribram, a short distance away. So Ceplecha and his

---

* Meteorite hunters apply the loose terms 'bolide' to any meteor with an apparent brightness of above around −14, and 'superbolide' for anything over about −17. That latter is about 100 times brighter than the full moon. Both the Pribram meteor and the Chelyabinsk meteor I mentioned earlier make the cut as superbolides.

colleagues got the word out to people living nearby: that bright flash was a meteorite and it might be lying in your back garden.

It was spring in Czechoslovakia and farmers were just beginning to prepare their soil for planting. One of the first things they did when the snow melted was to go through their fields and clear away all the stones. One farmer, a man named Vršecký who lived in the village of Luhy, not far from Pribram, had just completed this job. When he went back to his field a short while later, he noticed a large black rock. He knew it could not have been there before, realised there must be something special about it and took it home. Ceplecha's detective work led him to the man a few days later. It was there that he found a rock with jagged edges and a black crust, about the size of a cantaloupe melon, that weighed 4.5 kilograms. The team eventually recovered three other stones, each quite a bit smaller than the first. The last one was discovered in late August by a thirteen-year-old schoolboy.

These four stones were fragments of the Pribram meteorite. Ceplecha soon calculated that the rock that had entered the atmosphere would have weighed about 100 kilograms. He also used the trajectory caught on his films to show that it must have spiralled its way towards Earth, starting from the asteroid belt that lies between the planets Mars and Jupiter. Photographs from the time show Ceplecha, in a plaid shirt with slicked back hair, examining the rock with a magnifying glass. It was the first time anyone had been able to hold a meteorite in their hands and know the orbit it came from. Maria Gritsevich at the University of Helsinki, who specialises in modelling orbits, says the methods of doing this haven't changed much since the 1950s. She repeated the modelling of the Pribram meteorite's orbit in 2008 using state-of-the-art methods and found Ceplecha had pretty much nailed it. 'My opinion is that Zdeněk did really well for his time,' she says.

Ceplecha's retrieval of the Pribram meteorite fragments was a huge moment. But the rock itself turned out to be of a kind we were already quite familiar with. Scientists at the time already had plenty of meteorites in their collections. It was a pain not to know where they came from, but in the absence of that knowledge they could at least begin to put them into groups. As I learned in the basement of the Natural History Museum, there was one sort of meteorite that dominated all the others: the ordinary chondrite. About 87 per cent of the meteorites we find on Earth are of this kind. It turned out that the Pribram meteorite was an ordinary chondrite too.

As with so much in science, a single specimen is wonderful, but it doesn't enable you to easily draw any wider conclusions. But soon there would be more falls. Plenty of scientists were inspired by Ceplecha's success and tracking networks began to be founded in other parts of the world. In 1964, scientists in the US started the Prairie Network, a small tranche of cameras that watched the skies of the Midwest for incoming meteors. It wasn't until 1975 that they got lucky, with a meteorite known as Lost City. To the north, the Canadian Meteorite Observation and Recovery Network had also been started up. In 1977 it chalked up a find too and the Innisfree meteorite became the third space rock with a known orbit. Then in 1992 a really huge meteor seared through the skies above the US and sixteen separate video recordings of its fall, some made by members of the public, enabled scientists to calculate its orbit. One piece of the stone, known as the Peekskill meteorite and weighing twelve kilograms, crashed into the back of a cherry-red Chevrolet Malibu belonging to seventeen-year-old Michelle Knapp. She had recently purchased the car for $300 but quickly sold it on to a meteorite dealer for a cool $25,000. The car has since been displayed around the world.

Working with his colleague Pavel Spurný, Ceplecha expanded the camera network across the Czech Republic and into Germany. Over the next few decades it yielded two more meteorite falls and then in 2002 it bagged the most interesting yet: the Neuschwanstein meteorite. The curious thing about this stone is that its calculated orbit matched that of the Pribram meteorite almost exactly. It was an 'orbital twin', as Spurný puts it. This, he says, suggested that the Pribram and Neuschwanstein meteorites were part of the same stream of material caused by some epic collision between asteroids millions of years ago. Meteorite tracking was starting to give us neat insights.

Still, in the early years of the twenty-first century, meteoriticists were feeling both enthused and frustrated. It was incredible to have these stones, to be able to study them and know where they had come from, but, on the other hand, progress had been slow. In the roughly forty years since Ceplecha's first adventure in the strewn field, the collective efforts of science had managed to bag and assign orbits to a grand total of four more meteorites.

It was around this time that Matthieu Gounelle at the National Museum of Natural History in Paris had a bright idea. He had recently been studying the Orgueil meteorite, which fell at about 8 p.m. on a windless May evening in southwestern France in 1864. Gounelle's objective at the time was to compare the chemistry of the Orgueil rock with other, more modern, meteorites in order to answer some subtle questions about cosmochemistry. But as he worked, he became more and more entranced with the Orgueil stone.

It is a meteorite with a particularly storied history. Gounelle has called it 'one of the most-studied rocks of any kind'. Among other things, the meteorite was once the focus of an elaborate hoax. In 1964, researchers noticed a seed embedded beneath the surface of one fragment of the meteorite, which had ostensibly been kept in a sealed jar for the past 100 years. This generated much excitement, naturally. Could this be evidence of extraterrestrial life? Alas, the seed was eventually found to be that of a kind of rush, an Earthly plant, which had been glued in place and covered with coal dust. To this day, no one knows who carried out the deception.

What made Orgueil especially entrancing for Gounelle, though, was the incredible detail in which its fall was recorded. In France, it was the heyday of positivism, the ideology developed by the philosopher Auguste Comte, who entreated people to use logic and data to draw conclusions and to repudiate metaphysical speculation. It was the dawn of the age of natural history; science was fashionable and many educated, wealthy people were into collecting specimens and studying the natural world. This meant that when a flaming meteorite seared through the evening sky, lots of people recognised it as a scientifically interesting phenomenon and jumped at the chance to contribute their observations. Many wrote to the French Academy of Sciences with detailed notes on the time and the exact position of the fireball in the sky. They knew the constellations well, so they were able to give precise descriptions of the meteor's path. Today, with so much light pollution, few of us know our way around the night sky with any confidence. 'Nowadays this would be impossible,' says Gounelle. 'It tells us a lot about knowledge that we have lost.'

Gounelle wondered if he could reconstruct the Orgueil meteorite's orbit from the descriptions. He collected the observations and sent them to Pavel Spurný, who reconstructed the path of the meteorite and extrapolated it backwards to get an orbit. It wasn't the most accurate calculation ever made, because of the source material, but it was good enough to show that the meteorite's orbit was highly unusual. Its aphelion – the point of the orbit which is furthest from the sun – turned out to be far beyond Jupiter. 'It came from the outer solar system,' says Gounelle. 'And what would you call objects that live beyond the orbit of Jupiter? Comets.'

That seemed distinctly strange. We tend to think of small celestial objects as divided into two classes: asteroids and comets. Meteorites are rocks and generally they tend to come from the inner solar system. Comets, by contrast, generally contain a lot of ices – frozen water, methane and such like – and we tend to think of them as frigid bodies, orbiting much farther from the sun, way beyond Jupiter.

For a meteorite to have come from a comet might seem odd, then, because comets are mostly made of material that would sublime from a solid into gassy oblivion if they were to strike a fiery passage through Earth's atmosphere.

But thanks partly to Gounelle's work, it has become more broadly accepted that the distinction between comets and asteroids isn't clear-cut. 'This boundary is like a semantic boundary,' says Gounelle. 'There are ice-rich bodies, which form beyond Jupiter, and ice-poor bodies which formed within the orbit of Jupiter – and then you also have a continuum between these guys.'

All of this might make you wonder whether any of the other meteorites in our collections also come from what we might once have called comets. We can't be sure. There are no other meteorites with known orbits that extend beyond Jupiter (though a few come close), but there are a handful with similar kinds of composition to Orgueil, the precise origins of which we do not know.

That goes back to the fact that, in the mid-noughties, there were still only about ten meteorites of any kind with a known orbit. In 2005, Spurný and the British meteoriticist Phil Bland – whom we have already met in connection with the Chinguetti meteorite – decided to set up a new network of tracking cameras in a place they hoped would be easier to search and so yield a better crop of meteorites. With its straight roads that stretch on endlessly, wide skies and wombat burrows, Australia's Nullarbor plain is one of the world's most remote deserts. The idea was that Bland would lead a team based at Curtin University in Perth to track the trajectories of incoming fireballs, predict the strewn field and then drive out into the desert to find them. These would be epic expeditions, hundreds of miles from the nearest town. But in theory they had a much greater chance of finding the rocks out there, without any of the long grass or woods that make meteorite hunting so difficult in much of Europe. The Nullarbor plain has dark skies with no light pollution, minimal cloud cover, little vegetation and a pale whitish-orange bedrock against which the dark fusion crust of meteorites stands out – it's a fantastic place to hunt for fallen space rocks.

Bland's first hunt began on the night of 20 July 2007, when a bright white fireball ripped through Earth's atmosphere. Bland and his team traced the fall to a spot in the outback and eventually he and seven others set out in a truck and three cars carrying water and supplies for two weeks of camping. Ellie Sansom is now the manager of the project Bland started, known as the Desert Fireball Network or DFN. She wasn't on that first search, but she has been on several meteorite-hunting field trips. It's about a five-day drive to get out onto the Nullarbor plain, she says; three on roads and two following dirt tracks. 'Usually the last twenty kilometres is what we call bush-bashing, in other words driving out into the absolute middle of nowhere,' she says. The team get to the location, set up camp with 'swags', canvas tents with a mattress inside, and go on a long drive to find some firewood. The trip often lasts two weeks and the only link back to civilisation is a satellite phone.

What's it like out there? 'On my first trip I took a load of papers to read, thinking I'd have loads of down time,' says Sansom. But it turned out to be exhausting work. The job involves the researchers walking in a line formation across the desert, scanning their eyes over the ground for hours at a time (a safer option than using metal detectors, in case the meteorite doesn't contain iron). You need to search at the right time of day so that the sun isn't casting long shadows that could mask a meteorite. 'In summer, though, you have to have down time around mid-day,' says Sansom. 'That's obviously prime search time, but it can be more than forty-five degrees Celsius. It's just not safe to be wandering around in that heat.' The mental exhaustion gets to you too, so you need people who can make banter on these trips, says Sansom. 'One thing we do is pick a song that everyone knows and try to put meteorite lyrics to it.'*

It's not just meteorites they tend to find out there. There are also strange objects called tektites: jet-black droplets of glass. They were created hundreds of thousands of years ago when a gigantic meteorite smashed into the Earth, melting the ground and sending molten droplets of rock high into the atmosphere. 'These are basically melt-drops from this impact which have rained down across the globe,' says Sansom. 'In certain areas where the ground is really old, you can find loads of them. It's actually a good way of seeing who is concentrating – who finds ten a day versus who finds one a day.'

Back on that first expedition, Bland's team were lucky. They found a meteorite on their first day, within 100 metres

* During our interview, Sansom gives me a sense of what she means, singing out 'I came in like a fireball!' to the tune of Miley Cyrus's song *Wrecking Ball*. I am deeply sorry that the medium of the written word prevents me from reproducing this in its full glory.

of where they had predicted it to be from the trajectory calculations. Bland later found another chunk, making a total of 324 grams of rock by the end of the trip. They christened their find the Bunburra Rockhole meteorite, after a nearby cave. And again, the Bunburra rock proved the value of trajectory tracking. It turned out to belong in a group of meteorites called the eucrites, the members of which lack those enigmatic chondrules. But it was not just any old eucrite. Based on their chemistry, most eucrites are thought to originate from a large asteroid called Vesta that orbits in the asteroid belt, but the Bunburra Rockhole meteorite had an orbit almost entirely contained within Earth's, suggesting that it came from a different parent asteroid. If it wasn't for Bland's network it would still be lying undiscovered in the desert.

Since then, the DFN team have recovered three more meteorites. That's a considerable uptick in success rate compared with other networks. The team has managed to bag four 'falls' in about ten years – as many as the collective efforts of science had managed in the preceding fifty years. Now there is a different bottleneck: having estimated the locations of fifteen further meteorites, Bland's team are struggling to recruit enough people who know what a meteorite looks like and are willing to sleep in the middle of nowhere for weeks to bring them in. One way round this might be to use drones equipped with cameras and machine learning software to search for the meteorites. This is something Seamus Anderson at Curtin University in Perth, Australia, is working on. Together with Bland and others, he recently field-tested such a drone and showed that it could spot fake meteorites – essentially small stones painted black – though it also flagged a lot of stones and small holes made by animals as meteorite candidates. In future, maybe we won't need people to search for meteorites at all.

It may not be as easy to spot and find meteorites in places like the UK, with its cloudy skies and scrubby woodlands. But then again, the technology used to video the fireballs has improved and the methods for calculating where they will drop have become more precise. That's why, spurred on by the success of the DFN, there is now a movement to reinvigorate tracking networks in other places around the world. In 2017, Luke Daly, who worked with Bland in Australia, moved to the University of Glasgow. At the time, there was already a handful of organisations scanning the skies of the UK for meteors, mostly run by citizen scientists with cameras on their roofs. Aided and encouraged by one of them, Jim Rowe, Daly drew a handful of these organisations together into the UK Fireball Alliance, a project with the express aim of recovering falls in the UK. It was this effort that found the Winchcombe meteorite.

The story begins on a Sunday night in February 2021. The second wave of the Covid-19 pandemic was dying down in the UK, but the country was still under strict lockdown. The pubs were shut and so was almost everything else. Like most people, Áine O'Brien, a PhD student at the University of Glasgow, was at home, in her case idly browsing Twitter. The first hint that something was afoot was when she started seeing people posting unusual videos from their doorbell cameras. These showed a bright light streaking across the skies: a meteor. Even now, only a handful of meteorites are seen to fall – whether by cameras, or just by human eyes – and then be actually recovered around the world each year. In 2016, for example, there were only eleven. In the UK, the last time a fall was observed and recovered was in 1991.

This was the first time that Daly's Fireball Alliance network had picked up something that looked like it would drop a meteorite on land, but it couldn't have come at a worse time. No one involved was expecting to be allowed to go out and look for it because of the pandemic restrictions. Daly and Ashley King, a meteorite expert at the Natural History Museum in London, went on TV and radio news to ask local people to keep an eye out for unusual-looking rocks. At that stage it seemed there was little else they could do but get the word out and hope.

By Monday, the Wilcox family of Winchcombe, Gloucestershire – Rob, Cathryn and their daughter Hannah – had reported that a black smudgy rock had fallen on their driveway the night before. Hannah later told the BBC that she'd heard something fall but couldn't see what was going on in the dark. 'It was only the next morning when we went out that we saw it on the drive, [looking] a bit like a kind of splatter. And in all honesty, my original thought was: has someone been driving around the Cotswolds lobbing lumps of coal into people's gardens?'

Richard Greenwood at the Open University in Milton Keynes went to visit the Wilcox family. They had bagged up the meteorite just as King had suggested they should on the news. It looked like crumbly black charcoal, but for Greenwood it was the work of a moment to recognise that this was a special meteorite, probably a carbonaceous chondrite. As we saw when we encountered them in Costa Rica in Chapter Two, unlike most other rocks from space this kind of meteorite is not just a dry old stone but one crammed with water and complex carbon-based molecules, the kind of stuff that life is made from. 'I walk up and there is this country's most interesting meteorite in 400 years – and I was the first person that knew,' Greenwood told the BBC. 'It was very hard to contain the emotions.'

It was enough to make Daly start wondering if, despite the lockdown, he would be justified in organising a wider search for other fragments. By this time, Daly's old colleagues at Curtin University had calculated the area where fragments of the rock ought to have landed and narrowed the strewn field down to about five square kilometres. 'That's the kind of area that you can effectively search with say seven people in a week,' says Daly. He was itching to get down there. After some lengthy discussions with the university's administrators, and a rigorous risk assessment, the okay was given and he drove to Gloucestershire with his colleagues Lydia Hallis and Áine O'Brien, and his partner Mira Ihasz, along for the ride.

The four of them arrived in Gloucestershire and met up with another fifteen or so scientists from several other universities. They began their search together on a hilly common covered in gorse bushes and started walking around in long lines – socially distanced, of course. But there were plenty of people out walking their dogs and giving them funny looks – meeting up in groups was against the law at the time. So the scientists decided to start trying their luck on private land. It was a case of knocking on doors and asking people's permission. Almost without exception, people were interested and happy to let the scientists do their thing, so they got to it.

Only Daly, Katie Joy at the University of Manchester and one or two others had done meteorite searching of any kind before, so they gave the rest of the team a mini masterclass. Your instinct might be to look directly down, but it's better to look about four feet ahead and sweep your vision from left to right, explained Joy, as your eyes will naturally pick out contrasting colours. So the scientists formed lines and walked slowly across field after field. By the Friday evening they had been searching for two whole days and were starting to think they were not going to find anything. 'Gosh, the amount of time we spent staring at lumps of sheep poo, thinking it was a meteorite when it wasn't,' says O'Brien. 'There were a lot of laughs about the "meteo-wrongs" that we found.' Some scientists couldn't stay any longer, but Daly, O'Brien, Ihasz and a few others decided to try one more day.

The next morning, the team caught the owner of an estate and asked for permission to start searching the land. They picked a field and started looking. Ihasz was the only person there who wasn't a scientist. She had been at the hotel all week, working on her day job while Daly and the rest of the team were out searching, but because it was a Saturday she had joined in. She thought she'd seen something about eight times in the first hour, only for it to be – of course – sheep poo. Then, at exactly 9.39 a.m., she thought she had found a rock. She was nervous about stopping the line again for a meteo-wrong, but tentatively called out to Daly. He looked and immediately knew she was right. O'Brien started filming with her phone and recorded the moment they found the meteorite. She is screaming wildly; Daly and Ihasz are jumping up and down.

The scientists had a lot of paperwork to do that day, but they did stop briefly to celebrate. One of the people who lived nearby had been liaising with O'Brien and had sent her a message with a photo of a picnic table outside her cottage set with a bottle of whisky and six glasses. 'It would have been rude not to accept,' says O'Brien. 'So after a morning of chaos, we stopped for our lunch and we all had a wee dram of single malt, which was just gorgeous, and so poetic, being the Glasgow team.'

What do we gain from our efforts to track meteorites as they fall through the sky and then find them? In a sense, we're still at the beginning of this journey. What we'd like to do is chart how the different chemistries of meteorites relate to their orbits and see what this tells us about the history of the solar system. But there are still only a little over fifty meteorites in our collections with known orbits, and that isn't enough to give us a good sense of the lie of the land. Plus, it hasn't proved possible to trace meteorites' orbits all the way back to their parent asteroid. Meteorites are thought to have been cracked off their parent asteroid in some epic collision at some distant time in the past. From that point on, their orbits will have subtly changed over time until they eventually cross Earth's path. There are other ways of matching a meteorite to its parent asteroid, one of which involves actually visiting the asteroids with spacecraft, as we will see in the final chapter. But it's not possible to reconstruct the entirety of a meteorite's orbital history from the day it was born in perfect detail.

That might all seem rather deflating. Not so. One thing we get from tracking meteorites in this way is that we find them at all. It's not like, if we hadn't tracked the Winchcombe meteorite, we would have it anyway and just have known slightly less about it. By finding more meteorites, we also begin to find increasingly rare ones. The Winchcombe meteorite did indeed turn out to be a carbonaceous chondrite, which probably had its origins in much smaller, darker asteroids that formed further away from the sun. Because they are so small, their interiors wouldn't have been squeezed so much and so they didn't heat up on the inside. 'They never really got hot,' says Ashley King, at the Natural History Museum, who led the efforts to analyse the Winchcombe meteorite. 'They're really primitive, unmodified; they're basically the building blocks of our solar system.'

The carbonaceous chondrites are divided into eight sub-types depending on their chemical composition. The Winchcombe meteorite is part of the CM group, members of which contain somewhere around 14 per cent water by weight. They also contain a lot of carbon-based molecules, including things like amino acids, the sort of chemicals that all living things on Earth are made of. It's possible that this kind of rock is what delivered the water and organic chemicals needed for life to get started on the young Earth – an intriguing question that, again, we'll delve into in more detail later.

But out of the 500 or so CM carbonaceous chondrites we have in collections, nearly all are finds – in other words, meteorites that were picked up many years after they landed. This means the materials inside them will have changed. The Winchcombe meteorite is by far the most pristine CM carbonaceous chondrite sample we have. The Wilcox family had their specimen in a sealed bag less than twelve hours after it fell to Earth. 'Short of going to an asteroid in space and taking a sample from there, it's about as fresh as you can get,' says King.

What that gives us above all, says King, is confidence. It provides a sample that we really know is pristine and has barely been altered while on Earth. One of the ways in which we analyse other meteorites is by heating bits of them up and analysing the gasses that emanate from them. If you do this on old carbonaceous chondrite samples, you tend to get an awful lot more water bleeding off in the outer layers of the meteorite than you do when inner parts of it are tested. But that doesn't happen with the Winchcombe meteorite, which suggests that excess water in the older meteorite samples is Earth's atmosphere that has absorbed itself into the outer layers of the rock. Winchcombe, then, is like a trusted baseline. 'It helps us interpret what's terrestrial and what is extraterrestrial in our other samples,' says King.

Both fragments of the Winchcombe meteorite are now at the Natural History Museum. The one found by Ihasz is in the glass jar on public display in The Vault, but the fragment that landed on the Wilcox family driveway is kept behind the scenes, away from public view, so it can be analysed. In fact, I caught a glimpse of it that time I visited Martin

Suttle in the basement. The fragments of the Winchcombe meteorite were kept off to the side, in a clear plastic vacuum chamber on a lab bench: a humble place to store a sample of one of the world's most remarkable space rocks.

CHAPTER FOUR

The Mystery of the Missing Meteorites

I REMEMBER ONCE talking to my son about Antarctica and at some point in the conversation we scrambled upstairs to look at the globe in his bedroom. I picked up the globe and turned it upside down, expecting to point out the amorphous white blob that represents Antarctica – but we could barely see it at all. The plastic widget that fits the globe into its supporting arm almost completely obscured the continent. And that just about sums up the relationship that most of us have with this frigid land: out of sight and out of mind.

William Cassidy was an exception. He travelled the world studying impact craters – the gaping holes chiselled out of the ground when large meteorites or asteroids crash down. He had long been curious about Antarctica, though he had never been lucky enough to visit, but in the middle of his career he happened across a nugget of information that would set his life and the study of meteorites on a fresh course. Not only would he go to Antarctica but he would also make a discovery there that transformed the way scientists search for meteorites.

You'd be forgiven for feeling surprised at all this. Even getting to Antarctica is a long and challenging journey, and searching for meteorites there – thousands of kilometres from civilisation in the punishing cold – might sound slightly unhinged. But there are good reasons why this is the world's best place to hunt meteorites, reasons that Cassidy would play his part in revealing. Today, incredibly, two thirds of the meteorites in our collections come from Antarctica. Still, there are enduring puzzles that remain, not least the mystery of Antarctica's missing iron meteorites.

The story starts in August 1973, when Cassidy was in Davos, Switzerland, for a scientific conference. There, he heard a talk by chemists Makoto and Masako Shima, a Japanese husband-and-wife research team, who were describing their analysis of four meteorites. Cassidy attended because he had read that these meteorites had been found in Antarctica.

They had been recovered four years earlier by a Japanese geologist called Renji Naruse. He was part of a team of scientists who travelled to a remote spot in Antarctica called the Yamato Mountains, to measure fundamental things like the rate of the ice flow in the local glaciers and the kind of ice crystals found there. One day, Naruse noticed a rock lying on the ice that he recognised as a meteorite. He and his colleagues decided to take a break from their normal measurements and spend that day checking if there were any more. To their presumable surprise, they found a total of nine, four of which had ended up in the hands of the Shimas for chemical analysis.

Listening to the talk, Cassidy took in this backstory with at first only mild interest. As we know, to find several meteorites in proximity to one another isn't unusual, because of the way incoming stones often break into fragments in the atmosphere. But then Cassidy twigged something startling: the Shimas were saying that their meteorites were all different types. These were not part of the same parent meteor, but all came from separate falls. Cassidy knew that was hugely improbable. Only four meteorites had ever been found in Antarctica before – to find nine from separate falls in a single day's search was truly wild. 'That is when the lightbulb went on over my head,' Cassidy later wrote in his memoir. He began to think that, if the Shimas' story was true, there must be some mechanism concentrating meteorites in the Antarctic ice.

Cassidy was excited by the idea that Antarctica might be an unsuspected hotspot for meteorites and he wrote that same autumn to the National Science Foundation (NSF), a US body that funds scientific research, proposing that he take a trip to the icy continent to investigate further. It was turned down on the basis that Naruse's find could just be a fluke. Sending people to Antarctica is expensive, after all, and no one is going to shell out the cash lightly.

University professors may not always have the most glamorous lives, but they do have certain connections. In Cassidy's case there was a man by the name of Takeshi Nagata, who was a visiting professor at the University of Pittsburgh, where Cassidy worked. Nagata – known to Cassidy as 'Tak' – was the director of the Japanese National Institute for Polar Research, but he would periodically spend time in Pittsburgh. When he visited that autumn, Cassidy asked him about the amazing meteorite finds he had heard of. Nagata didn't actually know anything about them, but he immediately saw why Cassidy was interested and sent off a telegram – this was 1973 – to his colleagues to find out more. He ended up asking the field teams that were just about to head back to the Yamato Mountains for the next research field season to keep an eye out for more meteorites.

During the 1973–74 Antarctic summer, the Japanese research team found another twelve meteorites. It was encouraging but still not enough to persuade the NSF bosses to put their hands in their pockets for Cassidy to go. Then, during the field season of 1974–75, the Japanese researchers made an all-out effort to hunt down meteorites in the Yamato Mountains ice sheets – and came back with a haul of 663 stones. To put that in context, only about 2,000 meteorites had then been found across the whole surface of the planet in the previous 200 years. Well, that did it. Cassidy quickly got approval from the NSF for a trip to Antarctica to hunt meteorites.

We now know that Antarctica's glaciers and ice sheets act a bit like one of those conveyor belts at a supermarket checkout. You gradually unload your shopping trolley onto the belt, which carries all the toothpaste, cauliflowers, pork chops and whatnot to the till. Like that rubber conveyor belt, Antarctica's glaciers are also moving, albeit very slowly – not more than a few metres a year. Anything that lands on their surface will be taken along for the ride. And this place is so remote and inhospitable that near enough the only things that do drop are meteorites. This means that in places where the ice flow is impeded or forced upwards – at the edge of mountain ranges, for example – meteorites get deposited and build up in large numbers. Cassidy didn't know all of this for a fact back in the seventies, but he alighted on roughly this hypothesis.

He didn't want to go hunting at the Yamato Mountains as the Japanese researchers had done, not least because it was incredibly difficult to get there. Antarctica is vast, much larger than the continental United States, and visualising its geography isn't easy, because the compass directions aren't much help (all directions are 'north'). But try thinking of it as a blob of ice in the middle of a circle of ocean, with the tip of South Africa at the top, New Zealand to the bottom right and South America to the bottom left. As a US researcher, Cassidy would be based at the McMurdo research station, on the side of Antarctica nearest to New Zealand. From there, the South Pole was a three-hour flight away and the Yamato Mountains almost twice as far again, on roughly the opposite side of Antarctica. (There's a sketch map of meteorite-related Antarctic geography on page 96.) Cassidy was going to need something a little more accessible, ideally within a helicopter ride of McMurdo station.

He became a regular visitor to the archives of the U.S. Geological Survey at their office in Reston, Virginia. Here they had detailed satellite and aerial photos of Antarctica. Cassidy found there were clear patterns in the ice – trails of rubble called moraines left behind by a glacier. He realised that he could follow these tracks on the photographs, as they merged and skirted obstacles. He could also see how the flow of ice sped up and slowed down, because fast-moving ice tends to form more cracks and crevasses. There were plenty of patches where the ice had to split and divert around a string of peaks – but in the middle it was trapped. Here the ice flow stops and meteorites would not be swept away. He found a number of these areas close to McMurdo and these became his first targets.

Along with his newfound collaborator Ed Olsen, Cassidy made the first leg of his journey, to New Zealand, in November 1976. The NSF flew all their scientists to Antarctica via New Zealand in a cargo plane. Cassidy did not enjoy the experience as the plane wasn't really designed for carrying people. In a letter home, he described the journey as like 'flying sitting backwards in an enlarged garbage can [...] while maniacs outside were beating the can with sticks'.

Cassidy took a few days to relax in Christchurch, New Zealand, before continuing his journey. Walking around the city, he came across a statue showing a man wrapped in bulky, old-fashioned cold weather clothing, holding a staff and gazing fixedly into the distance. The statue is of Captain Robert Falcon Scott, the British naval officer who led an expedition that reached the geographic South Pole in January 1912. They got there just five weeks after the Pole had been reached for the first time by the Norwegian Roald Amundsen and his team. On the way back, Scott and his three companions were overcome by bad weather, ran out of food and died. Scott is often seen as a heroic figure, and while Cassidy acknowledged this side of the man he also wrote that Scott had 'seriously overreached his abilities'. On this and every subsequent occasion on which he visited Christchurch, Cassidy would seek out the statue and stare at the figure of Scott, and then he would sit down somewhere quietly and firmly resolve to be more careful in Antarctica than Scott had been.

Soon Cassidy was on a plane jetting away from the sunshine and chatter of New Zealand and towards the frosted emptiness of Antarctica. The continent's weather is notoriously unpredictable, and is occasionally too awful for an aircraft to even attempt landing. Because they have only a limited amount of fuel, all flights to Antarctica cross a point roughly halfway to their destination where they must double-check whether they need to turn back; keep going beyond that point and there is no option but to complete the journey, come what may. Cassidy's heart fluttered as the pilots checked and rechecked the weather forecasts. Fortunately, on his first trip everything went smoothly. Staring out of the window, he saw peaks and valleys blanketed by snow, so that all the sharp rock edges had been rounded off. 'The scene looked more lunar than terrestrial,' he wrote.

McMurdo base, where Cassidy would soon arrive, lies on Ross Island just off the coast of Antarctica. The island also boasts two volcanoes, Mount Terror and Mount Erebus, and is covered in a thick layer of black ash from prehistoric eruptions, which is visible in the summer months when the snow has partly melted. Together with the industrial buildings, this coal dust-coloured dusting gave the place a dismal look to Cassidy's eyes.

When Olsen and Cassidy got settled in at the base, they discovered that 'Tak' Nagata and two of his colleagues from Japan were already there and they were also planning to go meteorite hunting on this side of the continent, rather than continue in the Yamato Mountains. Cassidy wrote that he was 'shocked' and disturbed at the idea of two teams from different nations competing with each other. According to his memoir, the chief scientist at the base, a man named Duwayne Anderson, had concerns too. He persuaded Cassidy and Olsen to team up with one of Nagata's men, Keizo Yanai, and work together. Cassidy was happy with that as Yanai had plenty of experience in Antarctica, having been part of missions to the far-flung Yamato Mountains. He would be a valuable addition to the team.

To hunt meteorites, the trio needed to venture onto the mainland ice sheets, so Olsen, Yanai and Cassidy were flown up to their chosen field site, where they unloaded their camping equipment with no trouble. Then the pilot, a veteran of the Vietnam war, said they had some extra time to explore and so the scientists asked if he could briefly fly them to the planned site of their second camp, to double-check if it was going to be suitable.

Then it all kicked off. When they touched down and got out, Yanai spotted a meteorite within moments. Olsen and Cassidy were ecstatic and started photographing it. Meanwhile, Yanai had started scanning the area with his binoculars – and then he abruptly ran off. Olsen and Cassidy, bewildered, began to run after him. The pilot, still in the helicopter, followed too. 'In silhouette, it would have been a memorable, if puzzling tableau,' Cassidy later wrote. 'A figure running at top speed over the ice, two more people chasing him and a helicopter skimming along behind.' As you might have guessed, Yanai had spotted another meteorite. They had found their first two within twenty minutes of being in the field.

It wasn't all such plain sailing, but over the next few weeks Cassidy and his colleagues certainly demonstrated that large numbers of meteorites could be recovered from Antarctica. One day, they arrived at what they thought was a typical moraine but it turned out to be composed exclusively of meteorites, some thirty-four in total, collectively weighing more than 400 kilograms.

From there, the project went from strength to strength and became known as the Antarctic Search for Meteorites (ANSMET), with Cassidy leading teams onto the ice during each southern hemisphere summer. In 1980, he hired a mountaineer called John Schutt to act as an expert field guide and Schutt went on to work with the project for almost the next forty years, finally retiring from field duties in 2019. Schutt remembers Cassidy, who died in 2020 aged ninety-two, as a big man with a slow voice; he was kind, insightful, a wonderful writer and a teller of awful jokes.*

Cassidy was also a 'champion sleeper', says Schutt. In Antarctica, you frequently have 'tent days' when bad weather prevents you from going outside. Cassidy would sleep through most of these. 'He had the ability to sleep anywhere, any time, and in any position,' says Schutt. Once, Cassidy was asleep when Schutt heard his muffled voice croaking 'help, help' from deep within his sleeping bag. The noise soon stopped and Schutt ignored it, but he learned later that Cassidy had got himself so relaxed that he had been physically unable to move. He had evidently

* One of his favourites, apparently, was: 'How do you define a gentleman? It's someone who knows how to play the bagpipes – and doesn't.' Perhaps the best that can be said is that it might have been funny in the seventies.

cried out for assistance – and then, having resigned himself to his fate, gone back to sleep. Schutt found it all extremely funny, though he admits he can identify with Cassidy a little more now he is in his seventies and sometimes naps in the afternoons.

At the end of that first season, all the meteorite finds were packed into crates, nailed shut and kept at the McMurdo base, with half to be shipped to the US and the other half to Japan. When Yanai received his delivery, however, he realised that one box was missing, the one with the thirty-four fragments found all at once. He raised the alarm, and the box was eventually found, hidden in a dark corner at the base. But when Yanai eventually received it, he confirmed that several specimens were missing. There was only one conclusion: someone at McMurdo had broken into the box and stolen a few meteorites, perhaps as a memento. The identity of the thief was never discovered.* 'Perhaps today those meteorites are sitting on someone's shelf or in a rock garden or something,' says Schutt.

The ANSMET project soon went on to shed light on an unusual and puzzling group of meteorites. The first of these fell in a place called Chassigny, France, in 1815 and another fell in Sherghati, India, in 1865. Then a shower of at least forty stones fell in 1911 at El Nakhla El Baharia in

* The truth about what really happened during this theft does seem to have been lost for good. When I asked James Karner, the current director of ANSMET, about the incident, he told me he had no knowledge of it. I also asked John Schutt. He did know of the incident and confirmed that to his knowledge the thief had never been identified.

Egypt. As researchers later began classifying meteorites into groups with similar characteristics, they realised that the stones from these three meteorite falls were all similar in composition. Gradually other specimens were found that also belonged to the same group. They became known as the Shergottites, Nakhlites and Chassignites and collectively as the SNC (pronounced 'snick') meteorites.

These stones were strange for two reasons. First, scientists knew they were igneous rocks, that is, rocks that have solidified from lava or magma. Second, the rocks had a very young crystallisation age, meaning they had solidified relatively recently. For nearly all meteorites, this age comes out at around 4.5 billion years, meaning they date from the solar system's first flush of youth, but the SNCs were only 1.3 billion years old. To space scientists, this suggested that these rocks must have formed in some ancient extraterrestrial volcano. That near enough ruled out the asteroid belt as the source because asteroids are almost never big enough to sustain volcanism.

So where had they come from? Discussions on this were simmering in the 1970s as Cassidy's ANSMET programme began to get into its stride. Scientists narrowed it down to just a few possible suspects – Mercury, Venus, Mars and Jupiter's moon Io – which all had volcanoes that were active within the requisite time frame. But no one could agree on which was the true source of the SNCs.

In January 1980, Cassidy and his colleagues were exploring an area known as Elephant Moraine, so named because the outline of the strewn rocks looks like an elephant from the air. One day they found a whopper of a meteorite weighing almost eight kilograms, which received the name EETA 79001. They could see this was an igneous rock from the get-go, so they knew it was unusual and it was therefore the first specimen to be analysed once back in a laboratory.

It turned out to be a member of the SNC group. The strongest hypothesis is that it formed during an apocalyptic moment when a large asteroid crashed into some planetary surface. The impact created such a powerful bang that chunks of the planet's surface were thrown into space and, millions of years later, collided with the Earth. In the moment at which these chunks were created, compressed air from the planet's atmosphere was injected into the rock by the impact shockwave, whereupon it was trapped inside molten droplets of material as they cooled into beads of glass within the rock. When EETA 79001 was analysed, these beads of glass were picked out and melted so that those atmospheric gasses could escape their billion-year imprisonment. Researchers measured the composition of those gasses and found it to be an exact match for the atmosphere of Mars, as established by NASA's Viking landers a few years earlier. It was proof that the SNC meteorites were pieces of Mars.*

* There's a story that one of the original Nakhlite stones that fell in Egypt in 1911 hit and killed a dog. If that were true, it would make the animal singularly unlucky. This meteorite would have been blasted off the surface of Mars 11.5 million years ago, a time before modern domestic dogs had even evolved as a distinct species. It then circulated in space until colliding with that one particular dog. It seems, however, that this story is apocryphal – or at least, there's no compelling evidence to prove this meteorite fall really sent a pooch to meet its maker.

Today, ANSMET is still going strong. Cassidy retired in 1995 and a scientist named Ralph Harvey at Case Western Reserve University took over running the project. In 2016, James Karner at the University of Utah became the co-leader and the expectation is that he will shortly take over from Harvey. When I ask Karner about how he got into meteorite hunting, I am expecting to hear details about his scientific career. But he begins by telling me a story about how, growing up in North Dakota, he and his friends would go to a nearby golf course and spend their time hunting lost balls – looking along the streets, among the bushes and in the woods – and how he loved doing it. For all the serious science involved in ANSMET, part of Karner's motivation is the primal thrill of discovery. 'I think humans as a species just have this hunting and gathering drive within them,' he says.

'Everyone has their contradictions, don't they,' Geoff Evatt tells me. 'Well, I've got some very large contradictions.' Sitting in his office in the mathematics department at the University of Manchester, I've been asking him how a mathematician got interested in meteorite hunting – an interest that eventually took him to Antarctica, too. And as he tells me the story, I begin to understand a little of what makes him tick. On the one hand, he loves to wrestle with the details of complicated problems and understand them at a precise level with mathematics. But he can't do that all the time – he couldn't cope, he says. That is why he has always tried to find ways to escape from rooms like the one we're sitting in – with its harsh strip lights and whiteboards – and

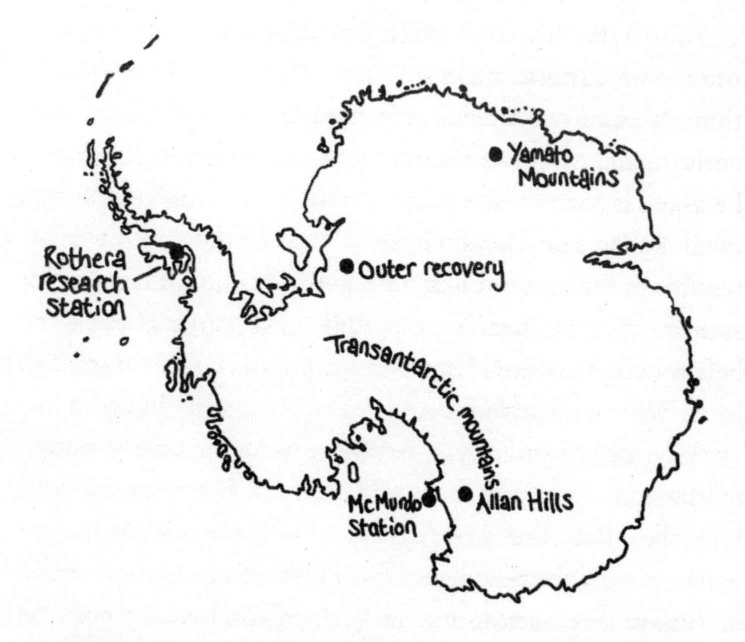

Meteorites on ice: the first Antarctic meteorites were a surprise discovery made by Japanese researchers in the Yamato Mountains. William Cassidy's expeditions operated out of the US McMurdo station and focused on the Transantarctic Mountains, including the Alan Hills. The more recent search for the missing iron meteorites, led by Katie Joy and Geoff Evatt, began on the other side of the continent, at the Rothera Research Station, and searched the Outer Recovery Ice Fields, among other zones.

get out into the great outdoors. Our meeting had to be in the first half of the week, because on Thursdays and Fridays Evatt is mostly to be found, chainsaw in hand, somewhere in the 120-acre expanse of Sunart Fields, a rewilding project in the nearby Peak District that he owns and manages with his wife.

Evatt talks fast and moves around a lot as he does so, waving his arms to make a point. When I met him, everything he told me seemed to be explained with unbridled enthusiasm. My favourite moment came after lunch, when he asked if I would like a chamomile tea to wash down the meal. When I said 'Yes, please', I was a fan of chamomile tea, his enthusiasm almost boiled over as he told me quite seriously that he often drinks seven or eight cups of the stuff before bedtime. Geoff Evatt is not a man who does anything by halves – even herbal tea.

Around a decade ago, Evatt was mostly studying the mathematical properties of glaciers, trying to understand how they flow and move. One of his favourite ways to work on such problems was to put out the call to a bunch of like-minded academics and organise a workshop in the middle of nowhere. They would all escape to a pub or hotel and spend a few days discussing some arcane aspect of ice dynamics. Evatt organised one such workshop in the Dolomite Mountains in 2012. On the way home, he and a few others were sitting in a café at Munich airport when another glaciologist wondered aloud to Evatt if these ideas about ice flow could be applied to meteorites.

William Cassidy's work had established Antarctica as the world's premier meteorite hunting destination, with more than 60 per cent of the world's meteorites coming from the continent. Yet, as Evatt learned as he began to think about meteorites and ice, there had always been a strange bias to the stones that were recovered out there.

As we know, meteorites can be split into three groups: stony, irons and stony-irons. If you look at the world collections of meteorites, the vast majority are stony meteorites; only 5.5 per cent are either stony-iron or iron. But here's the weird thing: if you look just at the meteorites recovered from Antarctica, only 0.7 per cent are iron or stony-iron.* That's not just a small discrepancy. It means the chances of finding an iron-based meteorite are almost ten times lower in Antarctica than they are anywhere else.

And iron-containing meteorites are worth having, not least because they are useful for studying the way planets form their cores. Earth has a molten iron core, the sloshing of which produces the magnetic field that protects our planet from harmful radiation in space (and produces the Northern and Southern Lights). When planets began to form, though, the iron would have been distributed through their rocks and it would have gradually sunk into the middle of them as they grew larger. It's thought that iron-containing meteorites might be fragments of planets that were part way through this core-forming process, but were smashed to pieces before they became big enough to endure.

Anyway, learning about the missing irons planted a seed in Evatt's mind and a while later he organised another of his workshops, this time at a countryside pub near Manchester, to talk about meteorites and ice. Several people came to give talks, including Katie Joy, whom we met briefly in the previous chapter. She and Evatt had first met through some

* These statistics are from 2020. The exact numbers tend to change ever so slightly each year as more meteorites are recovered both from Antarctica and around the world. However, the fact that you are about ten times more likely to find an iron-containing meteorite outside Antarctica has remained broadly true over the years.

mutual friends on a rock-climbing holiday in Spain. She had recently moved to work at the University of Manchester too and, as an expert on meteorites who had already been out to Antarctica hunting the things, it was obvious that she should join the workshop.

It was at the pub workshop that Evatt and Joy began to discuss a delightfully simple solution to the problem of Antarctica's missing meteorites. Iron meteorites are typically dark or black in colour and so absorb more warmth from sunlight than the lighter-coloured stony rocks. What if the iron stones were warming up to the point where they would actually melt the glacial ice beneath them and sink into it, perhaps even descending far enough so as to be hidden below the surface? It was just a hypothesis, but it would explain a lot, and it was so simple that it had to be worth testing.

Evatt worked out the mathematical model for how meteorites would absorb sunlight and the numbers seemed to check out. To test the idea, they would need some real meteorites, a block of ice, a freezer and a lamp. That led them to Andrew Smedley, another University of Manchester researcher, who is an expert on sunlight and the way its particular wavelengths of light affect materials. Together with some students, Smedley, Evatt and Joy dreamed up an experiment. They froze two roughly spherical and equally sized meteorites – one iron, one stony – in cubes of ice, specially prepared to contain no air bubbles, like real glacier ice. Then they stuck these in a huge walk-in freezer down the corridor from Smedley's office and shone a special lamp on the set-up to mimic the spectrum of light that comes from real sunlight.

They found that both meteorites sank. But the iron meteorite descended at 2.4 millimetres per hour, almost twice as quickly as the stony rock. We know that in places where ice flow is impeded by mountain bedrock, meteorites buried in the ice are forced upwards. According to the calculations, this meant that as the iron meteorites get close to the surface of the ice and begin to absorb sunlight, they could plausibly begin to sink into the ice faster than the moving glacier could push them upwards. (The stony meteorites are subjected to that downward sinking drive too, but much less strongly, so that the upwelling pressure wins out and they are slowly thrust to the surface.) The implication, the team thought, was that there could be a layer of iron meteorites hidden just beneath the surface of the ice sheets of Antarctica.*

It was a tantalising prospect. Joy and Evatt began to think about putting together a proposal to hunt for these concealed treasures, but it was not straightforward. In the UK, scientists often apply to research councils to fund projects out of pots of cash supplied by the government, but the idea of a wild expedition to track down space rocks

* Global heating will also have an impact on the rates at which meteorites sink into the ice. At the moment, researchers retrieve around 1,000 meteorites from Antarctica each year, while estimates suggest there are between 300,000 and 850,000 still exposed on the surface of the ice, waiting to be found. In an analysis published in 2024, Katie Joy and her colleagues looked at how increasing temperatures will affect the remaining rocks. Regardless of whether they are iron or metal, they will inevitably be heated to higher temperatures, melting and sinking into the underlying glaciers more quickly. Joy's study concluded that at least 24 per cent of the remaining meteorites will be lost in this way by 2050, no matter what action humans take to curb climate change. If greenhouse gas emissions continue to rise, 76 per cent of Antarctica's meteorites could meet the same fate by the end of this century.

buried in polar glaciers didn't exactly fit their brief. In the end, Joy and Evatt turned to the Leverhulme Trust, a fund that is known for supporting original research ideas. They also needed support from the British Antarctic Survey (BAS) in order to get out to Antarctica and live and operate there safely. It took quite a while to cook up a proposal that would convince both parties that it was worth doing.

Part of the strategy was to make the project about more than just hunting the missing irons. It would also – like ANSMET – be about simply finding as many meteorites on the surface of the ice as possible, the first British expedition to do this. When I met Joy, she came across as being just as friendly as Evatt, but she weighed her words more carefully. She knew from the start that finding the lost meteorites was a big ask. 'My motivation was primarily about finding as many new meteorites as possible,' she told me. 'We might find the one Venusian or Mercurian meteorite – something that's never been found before – that's why we want as many as possible. The question about the statistics of iron meteorites is fascinating, but it's a bit different.'

The first big problem to solve was how to find a meteorite hidden under the ice. The team considered several options, including ground-penetrating radar, but soon fixed upon the idea of using metal detectors to scan for extraterrestrial treasure. To make the search efficient, however, they wouldn't do this with hand-held instruments. Instead, they would use skidoos to tow a large metal-detecting mat across the ice. By driving several of these customised skidoos in formation, they could systematically search large areas of ice for the missing meteorites.

It was a solid plan, but not an easy one to realise – and it required the help of a cadre of other people with different areas of expertise. The team began working with engineers to develop the metal-detecting rigs, but they had not bargained for the weird ways in which electronics begin to behave at very cold temperatures. This made the work exceedingly finicky. And that was before, in an expedition to the other end of the Earth in preparation for their Antarctic travels, they encountered the nuisance that is the Arctic fox.

In 2018, Evatt and Smedley travelled to Ny-Ålesund on the island of Spitsbergen in the Svalbard archipelago. The world's most northern settlement, it hosts the Arctic research stations of several countries, plus an annual jazz festival, the occasional ultra-marathon and a large population of polar bears. Getting there from the UK is far easier than going all the way to Antarctica, so Evatt and Smedley were visiting the UK Arctic Research Station, run by BAS, to field test their metal-detecting skidoos. After the mandatory rifle training – so researchers can protect themselves from polar bear attacks – they got down to it.

The team constructed a kind of runway along the snow marked out with poles and flags and then they buried some fake meteorites – essentially lumps of iron wrapped with red tape to make them more visible – at varying depths along its length. The idea was to drive the metal-detecting skidoos along and check that they could detect the lumps of iron. But the team wanted a way to mark the hidden meteorites just in case they couldn't be found again. 'You don't want to spray paint on the ice, you can't put in a flag, 'cos you are going to run over the flag,' says Smedley. 'So, somebody suggested, why don't we use hot chocolate powder? It will biodegrade.' It seemed the perfect solution. The only trouble was the Arctic foxes came along in the night, licked up the chocolate powder, and then felt the need to mark their territory at each chocolatey spot. It meant the team's efforts to recover the fake meteorites were up against not just finger-numbing cold and the threat of polar bears, but Arctic fox excrement too. 'Those animals are not as cute as they look,' Smedley told me, with only half a grin.

Having shown that the skidoos worked, it was finally time to go to Antarctica. At the end of 2018, Joy and Evatt made their way to the icy continent, approaching from roughly the opposite direction to the one Cassidy had taken all those years before. After a stopover in Chile, they flew to the Rothera Research Station, the main Antarctic base of the BAS. It is situated on the Antarctic peninsula, a sickle-shaped slice of land that sticks up towards Chile.

It is a place not quite like anywhere else. The views of ice, snow and mountains are astounding. And though it's an exotic destination, the people you get there are a real mix, says Joy. There are scientists, ex-military pilots, plumbers, carpenters, cleaners and many others, all young. When Joy and Evatt arrived, huge amounts of building work were underway to build a new wharf for the new research ship the RRS *Sir David Attenborough*, leaving the base feeling like a building site. As for the wildlife, not all of it was quite as majestic as Sir David's documentaries depict it. 'There are elephant seals everywhere,' says Joy. 'They sit there farting and burping all around the base. Geoff thought they were hilarious.'

In that first field season, Evatt and Joy began to test out their ideas. Evatt took one of their prototype skidoo metal detecting rigs to an isolated area near a fuel depot called SkyBlu in order to test it out in punishing Antarctic conditions for the first time. Meanwhile, Joy went deeper into the interior of Antarctica, passing majestic sastrugi – dune-like structures formed in the snow by the wind. Her aim was to scope out the best place to go hunting properly when they returned the following year. She settled on a place they later named the Outer Recovery Ice Fields and also found thirty-six space rocks as she went along.

The following year, the team got out onto the Outer Recovery Ice Fields and began their search in earnest. They drove their snowmobiles over the smooth ice, dragging the metal-detecting rigs behind them. One day a beep sounded in Katie Joy's earpiece and a light flashed on the display fixed to her skidoo handlebars, which meant the metal detector had found something. She dismounted – could this finally be it, one of the lost meteorites of Antarctica? But alas, it was just a screw shaken loose from the detecting rig. The hard, bumpy ice gave the detectors a real beating and this wasn't the first time that something fell off the rigs, only to provide a false alarm later. In the end, the team found a total of about 120 meteorites on their two trips. But every one of these was on the surface of the ice. They did not find a single morsel of meteorite under the ice.

The obvious question is why – and it's not easy to answer. It's worth saying, however, that the researchers managed to search systematically a total area of only about one square kilometre and their calculations suggested that only about one iron meteorite would lurk beneath every two square kilometres of ice. It's not beyond the realms of possibility, given such a sparse spread of samples, that they could have missed a meteorite, perhaps because it happened to be lying under a rough area of ice that the skidoos couldn't traverse.

So, is there a chance of anyone picking up the hunt for the missing irons? When I ask Jim Karner, who runs ANSMET, about this, he tells me that he has indeed thought about it and has had conversations with Joy in this vein. The trouble is, he sees ANSMET's role as a service providing meteorites to the global community of scientists to study and as such the important thing is to retrieve as many as possible. So, could he really justify spending time on a tricky quest to find the hidden irons? As we talk it over, I can see him screwing up his face, wrestling with the question. He could spend a week searching in the normal way and potentially find dozens of meteorites, or he could spend that time on a possibly fruitless search for the hidden irons. He says he'd love to have a go but can't honestly justify it and Evatt feels the same way. 'It is very frustrating,' he tells me. 'I'd love to go back. But it's the funding that's difficult.'

There was another important outcome of Evatt and Joy's project that was a resounding success, however. From the start, Evatt had realised that he and the team had an opportunity to answer a simple but surprisingly tricky question: what is the flux of meteorites falling onto Earth, or put another way, how many fall on our planet each year?

It seems like the sort of question we ought to be able to answer. You might think we could just find a large area, search it for meteorites and then extrapolate from there. But there are many reasons why that isn't easy. For one thing, searching for meteorites is really hard, as we saw in Chapter Three – it's only really in places like deserts that we can be reasonably confident of spotting the majority of stones. The only place that has been comprehensively searched for meteorites is the Nullarbor plain in Australia, which is where Phil Bland's Desert Fireball Network operates.

This one search is not enough, however, because you would expect the incoming flux of meteorites to vary depending on the latitude. To get your head around this, imagine a shower of meteorites hurtling towards Earth. Because of the curvature of our planet's surface, this shower would be concentrated on a smaller area if it hit Earth near the equator compared to if it hit the poles. This suggests that there ought to be a lower density of meteorites at northern latitudes. But there is a second factor at play too: gravity. Some meteorites are on a course to sail 'above' or 'below' the Earth and these might be bent towards it by gravity – and these stones would inevitably hit the poles. In other words, there are two effects that ought to change the flux of meteorites hitting the Earth at different latitudes – but they both push in opposite directions, and we don't know the strength of each effect or which one dominates. This means we really need a good, comprehensive search for meteorites, ideally at an array of different latitudes.

Antarctica is already, of course, one of the best-searched areas of the entire globe when it comes to meteorites, thanks to Cassidy's ANSMET project. But there's yet another problem: here, the surface of the planet is largely a moving glacier, so when we find a meteorite in one place we know that is not where it originally fell.

Evatt realised he had an opportunity to solve this whole problem – and further his quest to find the missing irons into the bargain. If he could model the movements of the Antarctic ice sheet accurately, he would get a sense of how meteorites were transported around the continent in detail. This would tell him where the richest hunting grounds should be – which would be helpful for his own specific mission to find the missing irons. But it would also enable him to take data from expeditions that had searched a precise area of the ice and work out where the recovered meteorites had actually fallen.

The modelling was deep mathematical work, but before Evatt ever went to Antarctica, he, Smedley and their colleagues worked through the details and came out with the flux of meteorites in Antarctica. By comparing this with the previous measurements from Australia, he saw that the flux is in fact higher at the equator.* And by adding up the calculated flux of meteorites at all latitudes, he could work out the global flux of meteorites: it came out as an average of 17,600 stones weighing more than fifty grams each year. It is a number that will help scientists studying meteorites for years to come.

Antarctica's missing meteorites are still missing. But it's hard to see this continent as anything other than an

* How much higher? Compared to the global average, the flux of incoming meteorites is 12 per cent higher at the equator and 27 per cent lower at the poles. It is ironic that Antarctica is the best place to hunt for meteorites in the world even though it actually gets the lowest flux overall.

incredible bonus for meteorite scientists. Fifty years ago, it would have seemed like an absurd notion to go meteorite hunting in this inhospitable place. Today, the number of space rocks found in Antarctica stands at more than 47,000, an astonishing number. And every year, as bitter winds blow over the blue ice fields, the glaciers slowly nudge more extraterrestrial treasures towards the surface.

CHAPTER FIVE

On the Rooftops

JUNE IS A lovely time of year in the village of Brevik, Norway. The odd fragment of birdsong sounds through the stillness. An occasional shout drifts up from a child playing down at the edge of the fjord. The scent of pine needles and flowers fills the humid air. Oslo is just a few kilometres away and people from the city keep summer cabins here. Many of them like to eat the strawberries farmed in a huge field on the outskirts, which ripen right about now.

Jon Larsen and his family bought a dishevelled cabin here on the cheap twenty-five years ago and nursed it back to health. There was no soil on the surrounding plot of land when they first arrived, just a bare rocky hillside, but now there's a lush garden. Inside, short flights of stairs lead up and down at all angles to comfortable rooms. And then there's the veranda that looks out over a wooded valley. It was when Larsen sat down here to have breakfast one morning that the second all-consuming passion of his life found him.

Larsen carefully wiped the table clean of the leaves and detritus that had fallen on it overnight and went back inside to get a punnet of strawberries. But when he came back, a speck of dust lay on the white tabletop – one that hadn't been there a moment ago. He was absolutely sure of it, having wiped the top down not a minute previously. It came to him quite clearly: that speck of dust must have come from above.

For some reason, this seemingly unimportant occurrence wedged itself into Larsen's brain like a splinter. He had heard of meteorites, of course, and once the thought that the speck of dust could have come from space had entered his head, he could not get it out. He spent the rest of that summer's day in 2009 reading on the internet and before nightfall he had learned a new word: 'micrometeorite'. He put the speck of dust in a matchbox and stowed it in his cabin.

If Larsen had asked a meteorite expert whether it would be possible to happen across a micrometeorite by chance like this – or even to find one by intent searching – they would have said no. You couldn't find one on your veranda and you couldn't find one anywhere else either.

All that makes it rather wonderful that Larsen decided to dedicate himself from that day onwards to finding tiny, shiny particles of space dust lying around on Earth. He was a jazz guitarist by profession, and, as he toured the cities of Europe with his band, he would often sneak off after a gig to collect a bag of dust from the street or a rooftop. And what did he find in those bags? Well, let's just say no one was expecting it.

At more than sixty tonnes, the Hoba meteorite is the largest known single piece of meteorite in the world. The hulking chunk of silvery metal still lies where it fell at some terrifying moment in the past, in northern Namibia. Large extraterrestrial boulders like that slam down on us hardly ever (thank goodness), but the smaller they get, the more common meteorites become.

It is hard to be precise about these things, as we saw from Geoff Evatt's work on the flux of meteorites, but it is thought that approximately two olive-sized meteorites fall on any given area the size of Wales each year. However, when you get down to tiny, dust-sized particles, researchers suspect that our planet is being constantly sprinkled with them. Some estimates put the amount at about six tonnes – a good truckload – per day. That means your average rooftop, especially if it's flat, probably has a micrometeorite or two on it. The only trouble is that these motes of extraterrestrial dust don't make so much as a ping when they hit a roof, far less a fireball in the night sky. They come down to join us all the time, but no one notices, and they are instantly inconspicuous because they are so very small.

As a young man in the 1970s, Larsen was given a classical guitar as a present from his grandmother. He started noodling around on it and fell in love with the instrument. The melodic chords, the ring of a vibrato, the perfect juxtaposition of notes in a scale. Then one day he was listening to the radio when he heard some guitar music that really spoke to him. More than anything, it was the tone that impressed him. The musician was a French man with Romani ancestry named Django Reinhardt. 'No other guitarist has ever come close to achieving Django's powerful yet sensitive tone,' Larsen once wrote.

Reinhardt had died more than twenty years earlier, but his music so moved Larsen that he decided he must get closer to him, at least mentally, so he decided to travel to Paris, where Reinhardt had lived, to see what he could uncover. There he managed to find one of Reinhardt's old bandmates, Matelo Ferré, one of the few people still alive who had known the guitarist. However, Ferré could speak only French, so Larsen got hold of a French dictionary and learned the language as fast as he could, just so he could speak to him.

Django lost all but two of the fingers on his fretting hand in a fire when he was young. This prompted him to develop his own unique way of playing the guitar. His remaining two fingers became extremely strong, which was what enabled him to extract such incredible tones from his guitar in the music he played with his jazz band, The Quintet of the Hot Club de France.

Larsen went on to found his own jazz quartet with three of his childhood friends in 1979. It was named Hot Club de Norvège in homage to Reinhardt's outfit, and over the next forty-odd years he would tour Norway and Europe putting on concerts. Larsen was successful enough as a musician and producer to make a living for himself, but he never became a global star. He is probably Norway's most famous jazz guitarist – though these days he is better known in Oslo for other reasons.

Larsen was still touring intermittently with his band back in 2009 when his life was interrupted by the curious incident of the speck on the breakfast table. As he searched for more information about micrometeorites, he was – though he didn't know it yet – getting sucked into a new obsession. There were people on the internet who claimed collecting urban micrometeorites was an easy hobby. All you do is place a magnet in the downpipe from your gutter, then any iron-containing meteoritic dust that falls on your roof will stick there as it gets washed away when it rains. But these people didn't present any convincing evidence that what they found truly was extraterrestrial. On the other hand, some experts – notably Matthew Genge at Imperial College London, who specialised in the study of micrometeorites – believed that finding space dust in a polluted urban environment was not feasible. These experts got their samples from pristine, unpolluted places – ice cores from Antarctica, for example – or even collected the stuff before it hit the ground, using specially modified aircraft.

A major part of the reason for the experts' scepticism about micrometorites goes back to the early twentieth century and a scientist named Harvey Nininger, who is often considered a father of meteoritics. In the early 1940s, Nininger went to a site in Arizona where a meteorite had fallen thirty years earlier. He reckoned that this fall ought to have created a ring of fine dust particles from the meteorite and these could still be lying around. Together with his wife and another scientist, he dragged a strong magnet around in the dust and collected up the magnetic particles. Then, using a microscope, he painstakingly sorted through what he had found and discovered three tiny, spherical particles with a blackened outer shell. That suggested these tiny specks had been heated up, presumably as they seared through Earth's atmosphere at terrific speed. Nininger was convinced these were genuine micrometeorites.

Since Nininger's day, our understanding has moved on. It is true to say that when large meteorites pass through the atmosphere they shed tiny particles, but today scientists call these 'ablation spherules' and they are not the same as micrometeorites. Micrometeorites are the leftover dust from the early years of the solar system's genesis, dust that never formed into anything larger, like an asteroid or comet. In that sense, the cosmic dust that Larsen was after was older and more pristine than larger meteorites. Just as meteorites are fragments of asteroids that didn't make it into planets, micrometeorites are leftover dust that never became asteroids. They are, rather, the leftover dust from the material that formed the sun – micrometeorites are literally star dust.

Be that as it may, over the next few years a trend grew up to copy Nininger and collect putative extraterrestrial dust with magnets all over the US. But Nininger and his followers were too hasty in another respect. In 1953, two researchers from Iowa published a paper looking at the density of the 'cosmic spherules' that had been found across the US and noted that people tended to find fewer the further they were from an urban area. They also made a careful study of magnetic iron particles of fly ash, a glassy pollutant produced in industrial plants burning coal, which of course tend to be sited in urban areas. It turned out that this looked uncannily like Nininger's micrometeorites; it was black, shiny, round and magnetic. And it was produced in incredible quantities all across the US.

And it's not just fly ash. Today, we know that many by-products of human activity seem almost predestined to mimic micrometeorites. There's the dust produced by factories and car tyres, for example. Then there's the process of making mineral wool, a material that is commonly manufactured around the world and used as insulation for homes. This involves heating up basalt and chalk rocks to the point at which they melt and then blowing gas through the liquid rock so it expands into fibres. This process also forms tiny, shiny black spheres that inevitably escape into the environment. Likewise, fireworks produce an array of tiny and beautifully coloured spheres that look decidedly unearthly to the untrained eye.

Until recently, all those arguments were persuasive. One could not reasonably conclude that tiny black particles found in the dust were from space when there was a much more probable explanation. Martin Suttle, whom I first met in the basement of the Natural History Museum, and who studied with Matthew Genge, recently wrote about the Nininger episode in an article about micrometeorites. 'Consequently, from the 1950s onward,' he said, 'the notion that micrometeorites could be recovered from urban environments was considered an urban myth, unsupported by scientific research.'

If Larsen wanted to prove otherwise, he would be fighting against the scientific orthodoxy. But he reasoned that he had a chance. There surely ought to be differences between micrometeorites and the dust produced by human activities. His first thought was to look at the pictures of micrometeorites recovered from the deep sea and Antarctica. He discovered that Matthew Genge had even published a paper classifying all the different types of these micrometeorites. Yet these specimens are all extremely old and weathered by Earth's elements and, Larsen thought, probably not a good guide to what a fresh particle of cosmic dust on an urban rooftop looks like. Working out what a fresh micrometeorite looked like and what distinguished it from the legion of other motes of dust and dirt was going to be absurdly difficult.

But Larsen was undeterred. He began in Brevik, walking along the sides of the roads through the hills and collecting sackfuls of dust. Then he would take them home and go through the detritus using a USB microscope plugged into his computer. Larsen had a hypothesis. He assumed that extraterrestrial dust must fall on the Earth more or less equally everywhere. On the other hand, the constituent components of terrestrial dust would vary depending on what was nearby. In a dust sample from an industrial city, for example, you are likely to get more particles created by human activities. Gradually, painstakingly, he could learn to identify them, because they would always crop up in higher concentrations in certain places, with more in towns than in the countryside.

It took Larsen years to do it, but he began to build a catalogue of what was in the dust in order to identify that which was not of this world. The most common type of micrometeorite, Larsen discovered, is a tiny black ball made of the mineral olivine. Close up, it looks like a cannon ball, but scarred with many lines left by the atmosphere whipping past it. These are known as barred olivine.

Then there are the so-called cryptocrystalline micrometeorites. These particles have been melted into a glassy droplet on their way to the Earth's surface, but as they cooled they have just begun to form crystals – meaning the atoms inside the stone have begun to line up in a neat order – but this hasn't spread through the whole sphere. These meteorites are particularly beautiful, coming in a range of colours and aerodynamic shapes. Some are brown or light blue ovals, while others are amorphous and black, their surfaces dotted with bumps, like a turtle's shell. Many have a droplet of shiny metal visible at one apex. This is caused by a deceleration effect, like when a car brakes sharply and the passengers are thrown forward. As the micrometeor pushed lower into the atmosphere, it will have been slowed by the air resistance, whereupon the metal, with its greater inertia, slams towards the front of the bead.

There are other types too, often distinguished by exactly how their minerals have behaved during their hot descent to Earth. And then there are the really rare ones. There are the scoriaceous micrometeorites, which look almost as if they have been blown up from the inside. We think that these particles contained tiny bubbles of water, which heated up and exploded as they came through the atmosphere, popping the micrometeorite open like a piece of popcorn. Because they are expanded in this way, they seem to fall more slowly and don't get fully melted as they descend to Earth. There are also particles called CAT spherules, which are enriched with calcium, aluminium and titanium and which we know from studies of cosmic dust in Antarctica to be pure white in colour. Larsen has not found one – yet.

Larsen's collection of what he believed to be extra-terrestrial dust was growing: by 2015 he had around 500 specimens. By this time, he had begun to write to Matthew Genge. Would he look at the finds and confirm whether they were from space? Genge initially found this annoying. 'To begin with, I just wanted to make him go away,' he says. He received emails from people claiming to have found space rocks in their gutters all the time, but they never had – and Genge didn't have the time to investigate all the claims. But Larsen was persistent and when Genge did eventually glance at the photographs, they looked interesting. Eventually, he was persuaded to analyse the chemical composition of the dust and, in 2017, the pair announced that they had indeed found the first urban micrometeorites.

I can't remember exactly when I first read about Larsen's work, but it immediately caught my attention. I'm a sucker for stories that take something we think is boring or well-understood and show how it is in fact deeply unexpected and fascinating. The idea that something as simple as the dust on our rooftops could be hiding incredible extra-terrestrial treasures captivated me. I had to try hunting them too.

I found a ladder and climbed up to reach my roof guttering and then I used a garden trowel to scrape its contents into plastic sandwich bags. The weather had been fine, so it was mostly dry, dusty stuff; old bird poo and a few clumps of moss that the magpies had broken off the roof tiles. It wasn't the most glamorous fifteen minutes of my life. My wife took a photo of me and sent it to some of our friends, who reacted with a mixture of polite interest and mild disgust – but that wasn't going to stop me now.

Larsen had written a short guidebook for micrometeorite hunters and I followed the instructions. I took my bags of roof muck to the bathroom, emptied them into a plastic bowl, added water and washing-up liquid and stirred. Then I picked out the stuff that floated and, once the solids had settled at the bottom, decanted the brown water. The organic matter and dirt tend to mostly dissolve in water, while rocks don't. So after repeating this a few times, I was left with a dish of clean, tiny rocks and sand.

The next step is to separate your rocks by size. By careful experimentation, Larsen has found that most micrometeorites are between 0.2 millimetres and 0.4 millimetres long. He suggests using scientific-grade precision sieves to isolate that fraction of the grains. But those are expensive, so I put my mixture through an old tea strainer to get rid of the larger chunks.

Finally, I picked up a powerful magnet I had bought online, covered it in a plastic bag, stirred it around in the dust and transferred the magnetic portion of the stuff into a white bowl. This is a crude tactic because not every micrometeorite is magnetic, but many are and it massively reduces the volume of dust you are working with.

That was the cleaning process done, but now the real work would begin. I bought a cheap USB microscope, attached it to my laptop and began going through my sample grain by grain on my desk in the spare room. It is a beguiling world down there. Magnified sixty times, some specks look like pieces of multicoloured popcorn or dark, spiky stars. Others are translucent gems of many colours; red, blue, brown and delicate green.

I knew what to look for, thanks to Larsen's guidebook, but the truth is, this stage nearly broke me. I spent at least seven hours lost in this world of dust. Nudging around the grains with a chopstick, I sometimes found a promising-looking speck only to lose it again. At other times my mind wandered and I realised I had been mindlessly browsing this endless microscopic labyrinth, having lost sight of what I was supposed to be looking for. On several evenings, my wife had to explain that there were more important things to do right now, like read our children a bedtime story. In the end, I had to admit defeat.

I couldn't completely let go of my desire to hunt micrometeorites though, so months later I flew to Oslo to learn from the master. Larsen met me at the airport. He has a thick head of white hair and a friendly face, he speaks with a lilting Norwegian accent and has a habit of saying 'mmmm' in an intense way. This and his broad smiles made him sound always enthusiastic, no matter the topic of conversation.

In the airport car park is our transportation, a large white van in which Larsen keeps his meteorite hunting equipment: ladders, heavy-duty plastic bags, spades and a high-visibility gilet. We start driving and Larsen says that because it is going to rain later – rain is not good for meteorite hunting – he wants to take me straight to a roof where we can get our hands dirty. 'I have a question for you though,' he says, his face taking on a serious expression for the first time. 'Are you scared of heights?'

We pull up at a seemingly deserted industrial building which Larsen has scoped out in advance and got permission to ascend. There are three sections to the flat roof, he explains, and he has done the first two already. We just need to get up there, shovel the dust and dirt into some bags and get down again. But we need to work quickly; it is already spitting slightly and if the dirt soaks up too much moisture it will become so heavy that it will be hard to lift.

I'm not sure what I had expected, but the way to the roof is a metal ladder screwed to the outside of the warehouse. It is, quite simply, very high – the equivalent of about four storeys – and it goes straight up. If I fall off, there is nothing to catch me. 'Are you ok to climb that?' Larsen asks. Thoughts of the travel insurance I didn't bother to take out are running through my head, but I'm itching to get up there. So up we go, Larsen in his high-vis vest, me making sure to grip each cold metal rung of the ladder tightly before reaching for the next.

The view of Oslo's industrial sprawl isn't much of a reward for our climb, but then that's not why we are here. The roof is flat and smooth, with a ridge running around the edge; it is almost like being in a shallow swimming pool that's been drained. That makes it a perfect place for Larsen's work because all the worldly – and otherworldly – detritus that falls here is blown into the corners and stays there.

'You take me to the nicest places, Jon,' I say.

'Well, to me, this is beauty,' he exclaims, gesturing to the roof and giving me a huge grin.

As the drizzle stiffens, we stride over to the part of the roof Larsen hasn't yet tackled and I hold open a large plastic sack while he gamely spades the dirt and muck in. We fill two sacks and then lug them back to the ladder with some difficulty.

The only reasonable way to get them down is to throw them over the edge of the building. Larsen explains that you ideally want the bag to not split open when it hits the ground, but, unsure how to avoid that, I just loft my bag over the side and hope for the best. Back on firm ground, we find that both bags have split catastrophically, so we spend ten minutes trying to ease another bag over the top of each of them, to stop the precious filth from escaping. We chuck our haul into the back of the van and collapse into the front seats, the rain now rapping hard on the windscreen. 'You know, in Norway, they call your senior years the "age of dust",' Larsen says, starting the engine. 'I think I have reached my age of dust.'

The next day, Larsen picked me up from my hotel and we drove to his summer cabin in Brevik for the next stage of our work. On the way, we talked about his past, seeking out Reinhardt's legacy, and his tireless meteorite-hunting work. As we wove through the outskirts of Oslo I asked him as politely as I could if he thought he had a bit of an obsessive personality. 'Oh, definitely,' he answered without hesitation. 'Yes. Guilty.'

The summer cabin overlooks a steep wooded valley and it really is idyllic: pine floors, clean air and stillness. Larsen assembles his buckets and a garden hose as he's going to do more or less what I did in my bathroom at home, but on a larger scale. First he empties the sackfuls of roof muck we collected yesterday into two huge buckets and begins filling them with water. Then he swirls the liquid around and decants it carefully onto patches of flowers all over the garden. I'm surrounded by lush, well-tended plants of all kinds vying for space. They have all been grown on soil taken from Oslo's rooftops and stripped of its micrometeorites; there can't be a garden quite like this anywhere else in the world.

Larsen doesn't want me to help him with this process. 'I have never even had someone watch me do this before,' he says. He tells me he loves being out here alone, working with the mud, and he often gets into a flow state. I watch him hosing and swirling, then sinking his hands into the mud to massage out the lumps. He picks out a tiny bird bone and throws it into a nearby patch of mint. Eventually, I decide to leave him to his own private joy and take a walk down to the fjord. When I get back, he has two large dishes of clean stones, ready to hunt through.

In a true 'here's one I made earlier' moment, Larsen has already searched through the sample of dust he found on our roof last week and has found a dozen or so micrometeorites. He's stuck these to a paper slide in readiness for the final stage of his operation: precision photography. And for that, we are going to visit his long-time collaborator.

Meeting Jan Braly Kihle is an interesting experience. Physically striking, he's a big man with a wide moustache and a penchant for hats and sequined shoes. He has survived cancer three times and his car licence plate reads 'GRAT3FUL'. Although a geologist by profession, he is also an expert on technology. He once invented a magnetic CD clamp that bends the disc by a small, precise amount to greatly reduce errors in playback. Larsen met him years ago and they struck up a friendship as they worked on a way to produce photographs of his rooftop finds.

Kihle greets us quietly and ushers us down to his basement. It is spotlessly clean and adorned with an eclectic array of objects he's collected over the years: animal skulls, ropes, a beaver skin jerkin, a *Thriller* CD case signed by Michael Jackson, a boomerang. Pride of place goes to a camera that Kihle and Larsen have carefully modified to snap micrometeorite portraits. Despite the tiny size of an individual grain, the camera has to focus so tightly to capture its details that the whole of one can't be photographed at once. Instead, Kihle has devised a clever system whereby the camera adjusts itself to take hundreds of shots of the meteorite where just a small slice is in focus, photographing each elevation in turn. Customised software then stitches the images together. Capturing the full likeness of each micrometeorite takes perhaps half an hour. As the machinery is going over the new finds, we retire to some sofas in a side room and chat. When I'm not here, the two of them often sit around watching their favourite comedy show: *Monty Python*.

As the evening progresses, Kihle's computer screen shows us a sequence of tiny wonders. We see some barred olivines and some glassy spherules, but towards the end of the evening a truly bizarre specimen appears. This one is the shape and colour of an aubergine, but half of its surface looks degraded, as though it were mud in a dried-up riverbed. What exactly is going on here we don't know for sure.

But Larsen has a working hypothesis. He thinks while this fellow was falling to Earth its outer edge solidified, shrank and cracked into pieces, while its interior remained liquid. That liquid then oozed up to fill in the fissures. He even has a special term for this appearance: 'We say it is crackylated', he explained. It had also been on the roof for quite a while and was badly weathered, which added to the overall effect. If Jon had waited another few weeks before visiting that roof, perhaps this odd cosmic vegetable would have disappeared, utterly destroyed by the elements.

A few weeks after I returned from Oslo, Larsen emailed me to say he had found something really special in that dust. It was a scoriaceous meteorite, the type that has not completely melted. Months later, he wrote again to update me on the full analysis. It turned out that he had found a total of 141 micrometeorites in the buckets he processed, making it one of his most successful ever field searches. I would dearly love to take a share of the credit for this, but in truth my skills as a micrometeorite hunter have been weighed and found wanting. I failed pathetically to find any stardust on my own lousy roof and the best that can be said of my adventures with Larsen is that I held the bin bags open for him and managed to avoid falling off that terrifying ladder.

These days, Larsen is rightly regarded as an expert on micrometeorites. He has published many books and lectured extensively on the topic. The scientific community now accepts that you can find micrometeorites on rooftops and credits Larsen with proving it. Since he blazed the trail, others have followed. Scott Peterson is a veteran of the US army, but has now retired from service and become a student of science. He read Larsen's books years ago and became hooked on micrometeorite hunting and these days he is an accomplished practitioner. I follow him on social media and enjoy the pictures he posts of micrometeorites. (Larsen tells me that he thinks Peterson's photos of micrometeorites have now surpassed those he takes with Kihle in quality.) Sometimes Peterson posts photos of the assorted dust grains and asks his followers to spot the meteorite. I never can.

Thilo Hasse at the Museum of Natural History in Berlin also got into hunting micrometeorites a few years ago. He gradually amassed a collection of more than 1,000, including about 350 from a single large roof in the German capital city. A few years ago, he contacted Suttle, who is now based at the Open University, and asked if he would be interested in working with him on some science.

Suttle was all for it. What primarily appealed to him was the prospect of estimating how much micrometeorite material Earth is doused with in a given time. We do already have constraints on this number, some of which are based on measurements taken from space. Suttle himself had also been involved in a study with Luigi Folco at the University of Pisa, Italy, which looked at cosmic dust that had accumulated in cracks in remote Antarctic mountains over millions of years. That study estimated that between 800 and 2,300 tonnes of cosmic dust fell to Earth every year. The trouble is that studies like this are brimming with uncertainty. You ideally need to know the area the micrometeorites have fallen over and the time window too. But those mountain cracks could have been filled with dust blown from all over the place. And we also don't know how long the cracks have been there – Suttle reckons perhaps two million years, but that's basically a guess.

The Berlin roof is different. In 2022, Hasse cleaned the roof to a spotless finish and then began revisiting it every month to collect the meteorites that had fallen in the interim. He did this for more than a year in order to get a much more accurate estimate of the dust flux, as he would know precisely the area of the roof and the time during which the meteorites had fallen. The results aren't yet published, but they should provide our most accurate measure yet of the flux of cosmic dust.

Meanwhile, Penny Wozniakiewicz and Matthias van Ginneken, both at the University of Kent, in Canterbury, UK, are hunting for micrometeorites on the roofs of cathedrals. Van Ginneken had been interested in urban micrometeorites for years and had been thinking about how to study them. One day he was in his flat in Canterbury and his gaze alighted on the city's famous cathedral, which he can see from his window. He thought it would be a good place to look because the roofs are so seldom visited by people. Having obtained permission, Wozniakiewicz and van Ginneken climbed up to the cathedral roof with special vacuum pumps strapped to their backs. 'It's great, we can suck up the dust very efficiently in no time,' van Ginneken says. 'We do look a bit like Ghostbusters,' adds Wozniakiewicz. Their plan is to search for dust on several different cathedral roofs across the UK and try to get more detailed information about how the volume and kind of cosmic dust varies by location.

There is always the possibility that one day we will find a micrometeorite that came from beyond our solar system – from interstellar space. It isn't as far-fetched as it might sound. Telescopes have detected two much larger interstellar interlopers – one rocky shard called 'Oumuamua shot through the solar system in 2017 and a comet-like object called 2I/Borisov did likewise the following year. We can tell from measuring their trajectories that neither of these were in orbit around the sun and so they must have swept in from interstellar space.

Physicist and astronomer Abraham 'Avi' Loeb at Harvard University thinks he knows of another interstellar traveller: a meteor that fell in 2014, which has been nicknamed IM1. Loeb reckons that this object was travelling so fast that it must have interstellar origins, like 'Oumuamua and Borisov. He and his colleagues traced the path of IM1 (short for 'interstellar meteor 1') and found it should have come down in the ocean near Papua New Guinea, so in 2023 Loeb led a marine expedition to the location and dragged a 'magnetic sled' over the seafloor. Using this, the team dredged up 700 spherules, and having analysed fifty-seven of these, announced that five had a bizarre chemical composition 'never seen before'. Specifically, levels of the elements beryllium, lanthanum and uranium in the particles were about 1,000 times higher than those found in typical ordinary chondrite meteorites. This, Loeb claimed in a paper uploaded to the online arXiv repository, supports the interstellar origin of IM1. It is, however, fair to say that most other astronomers are extremely sceptical of these claims. The idea that IM1 is interstellar is not unreasonable

The Wold Cottage meteorite – an ordinary chondrite, the most common type of meteorite – fell in Yorkshire, 1795. The fall was witnessed by a ploughman, and was pivotal in helping scientists discover that stones from space can plummet to Earth.

Chemical Engineer / Wikimedia

Sliced and viewed close up, chondrites display tiny, multicoloured specks called chondrules. These microscopic wonders are the first motes of solid material that formed in the ring of gas that encircled the sun before the planets formed.

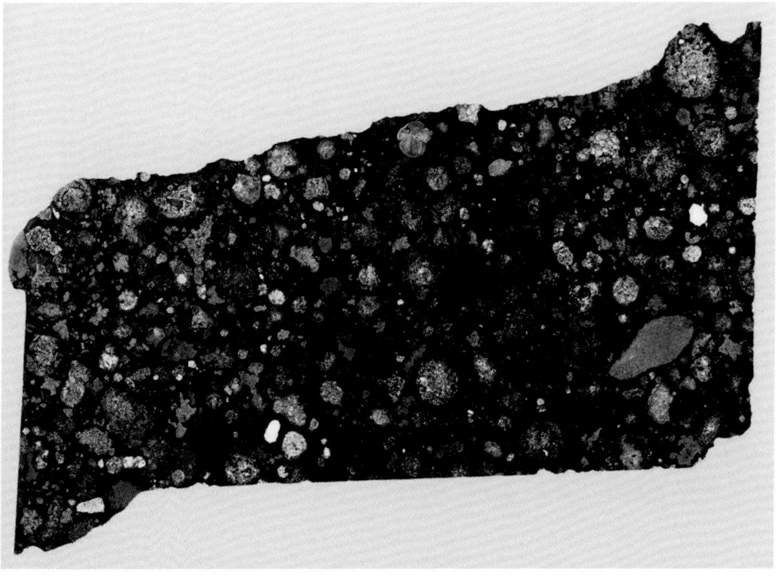

Solar Anamnesis / Wikimedia

Part of the Imilac meteorite, in the pallasites group. Made of metal and transparent olivine crystals, it is thought to have formed inside a fledgling planet at the boundary between its rocky upper portion and metallic core.

Steve Jurvetson / Wikimedia

Svend Buhl

German meteorite hunter Svend Buhl focuses on deserts and now leads a team called Meteorite Recon. Here he is in the Atacama, 2017, holding a freshly fallen meteorite named Los Vientos 188.

On his first meteorite-hunting trip, Buhl found only two space rocks, one part of the Taffassasset meteorite. It is a rare kind: a primitive achondrite. Meteorites are typically divided into those that have melted (achondrites) and those that have never melted (chondrites). This unusual specimen sits between the two.

Svend Buhl

Svend Buhl

Buhl's first expedition to the Ténéré desert in 2002. Some of Buhl's team (*left to right*: Souleymane Icha, Juergen Meineke, Rolf Poppinga, Jost Hecker) shelter from the midday heat.

Robert Ward

Robert Ward – the 'space cowboy' – is one of the world's most successful meteorite hunters. Once thrown into an Omani jail while searching, here he is during happier times in his meteorite collection hall in Arizona.

Pavel Spurny

Pavel Spurny

(*Left*) Zdeněk Ceplecha examines a fragment of the Příbram meteorite, the first meteorite to be recovered after its incoming trajectory was recorded, which he found in 1959.

(*Right*) Ceplecha holding the Luhy fragment of the same meteorite in 2009, fifty years after it landed.

Mira Ihasz / Luke Daly / Glasgow University

Initially mistaking it for a sheep dropping, Mira Ihasz found a fragment of the Winchcombe meteorite after a long and otherwise fruitless search.

NASA / JSC

William (Bill) Cassidy, who founded the Antarctic Search for Meteorites (ANSMET) programme, in Antarctica during the 1980–81 field season.

(*Left*) Katie Joy picks up a meteorite in Antarctica. It is important to avoid touching them – doing so could skew future chemical analysis.

Romain Tartèse / University of Manchester

NASA / Nina Lanza

(*Right*) Blackened meteorites stand out against the ice in Antarctica. Here is one in situ.

Geoff Evatt was part of the team hunting the missing iron meteorites of Antarctica using a metal-detecting rig towed by a skidoo. Here he is testing the equipment at the SkyBlu fuel depot.

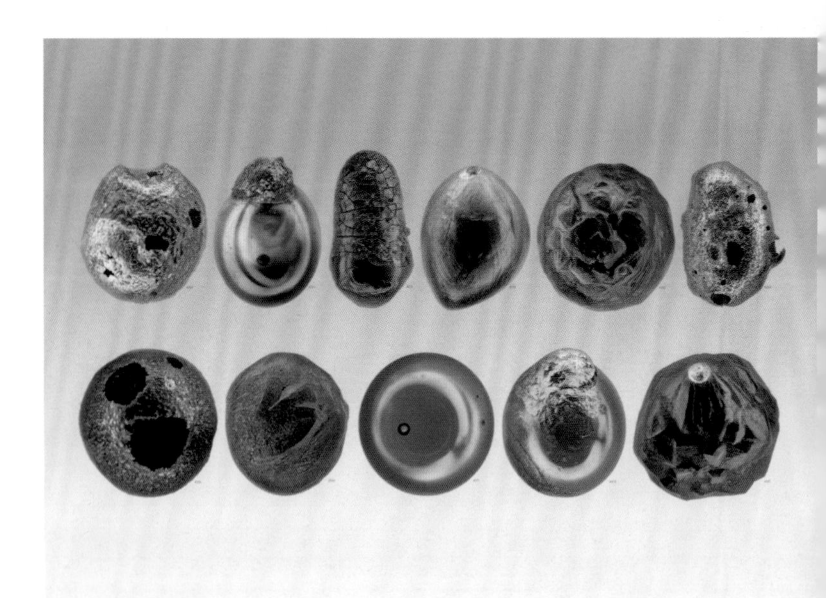

Jon Larsen's work on micrometeorites lives through the wondrous photographs he takes with Jan Braly Kihle. Here are eleven beautiful examples. Their shapes are created as they melt and recrystallise while falling through the atmosphere.

Jon Larsen

The author and Jon Larsen collecting dirt from a rooftop in Oslo on a rainy day.

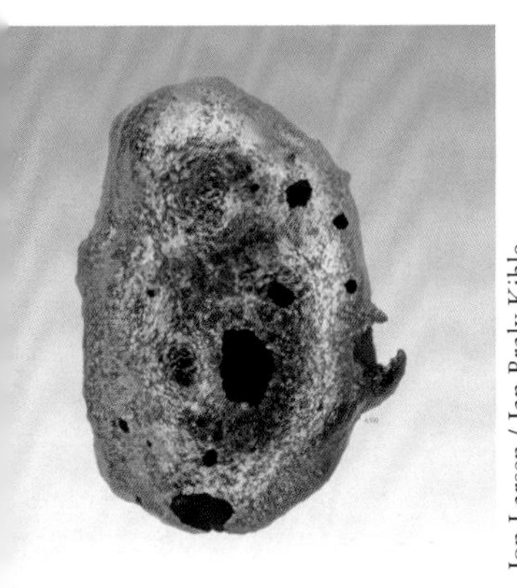

Jon Larsen / Jan Braly Kihle

Jon Larsen / Jan Braly Kihle

(*Left*) Some very rare micrometeorites don't appear to melt as they fall. Known as the scoriaceous micrometeorites, this one came from the batch of dust the author collected with Jon Larsen.

(*Right*) Some of Larsen's finds are what he calls 'crackylated'. He reckons these micrometeorites are beginning to be broken down by the elements. The dumbbell shape probably arises from an off-centre spin as it plummeted to Earth.

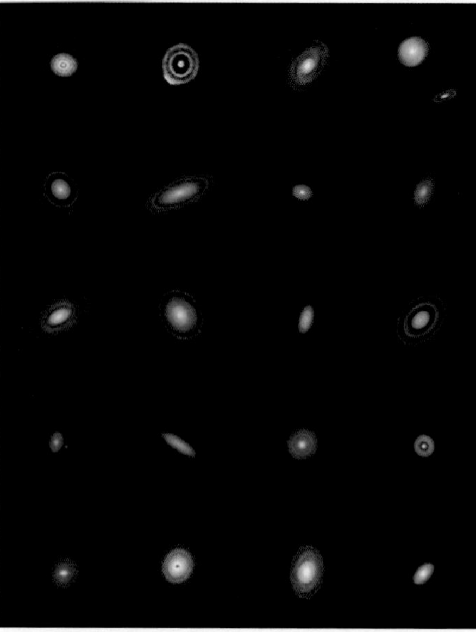
ALMA (ESO / NAOJ / NRAO), S. Andrews et al.; NRAO / AUI / NSF, S. Dagnello

Is the structure of our solar system – four small rocky planets followed by four gas giants – typical?

The ALMA array of radio telescopes in the Atacama desert provided scientists with key evidence, capturing these images of twenty nearby stars still surrounded by a disc of dust and gas from which planets will form. They come in a variety of patterns, suggesting that planetary systems' structures are probably extremely mixed.

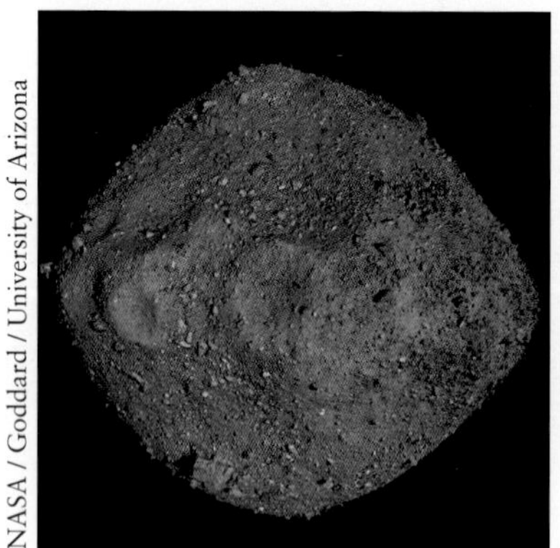
NASA / Goddard / University of Arizona

NASA / Erika Blumenfeld & Joseph Aebersold

(*Left*) The asteroid Bennu, shown as a mosaic of images taken by the OSIRIS-REx spacecraft.

(*Right*) A top-down view of the OSIRIS-REx Touch-and-Go Sample Acquisition Mechanism (TAGSAM). The dark material inside comes from Bennu. Analysis shows the asteroid may once have been part of a protoplanet covered in water and sprinkled with a range of the biomolecules needed for life.

in itself. However, it really isn't clear where it landed. Loeb and his colleagues used seismic data from a recording station near Papua New Guinea to track its inward path, but since then other scientists have suggested – rather amusingly – that the signals in question may have actually come from a passing truck. Moreover, the spherules Loeb found could have acquired their odd composition while being on the seafloor, or they could simply be a class of micrometeorite from inside the solar system that we haven't seen before. It's a cliché, but in science extraordinary claims require extraordinary evidence – and so far, that's lacking. Still, the search for micrometeorites will go on and who knows what we might find in future.

Larsen has now been searching for cosmic dust for fifteen years. But while he single-handedly revived the study of urban cosmic dust, his own future in the field is now tinged with uncertainty – not because of any waning enthusiasm on his part, but because the hunt costs money and no one pays him. When I was in Oslo, he told me that he was putting his entire micrometeorite collection up for sale. A few weeks later, I heard, with some relief, that he was putting that off while he first tried out Patreon, an online service where anyone can sign up to support a person's work with a monthly payment. As I was putting the finishing touches to this chapter, I checked the page and found he has fourteen supporters. While we were driving around Oslo, I asked him if he thought he might have to find paid employment at some point. 'No,' he replied. 'I have two kidneys.'

By the way, you might be wondering what happened to that speck of dust that fell on the breakfast table, the one that started it all. Larsen put it in a matchbox for safekeeping, but when he went to look for it later, he couldn't find it. Which means that this micrometeorite – if that is indeed what it was – is now lost and gone, adrift amid a vast ocean of uninteresting specks of dust, just like virtually all the others that have ever fallen on this planet.

PART TWO

The Secrets

CHAPTER SIX

An Unexpected History of the Solar System

WHENEVER I WALK over to the tiny kitchen in *New Scientist*'s office to make a cup of coffee, I pass a wall adorned with majestic images of deep space. You know the sort I mean. Think billowing clouds of amber and russet gas set against a backdrop of inky space and pink, twinkling stars. These particular pictures have even been etched into glass and there are lights inside the frame so the stars actually shine. They are undeniably awesome, in the true sense of the word, and yet, since I started researching meteorites, I always look at these images with narrowed eyes. In many cases, these kinds of images are, if not exactly a lie, then at least misleading. Usually, they depict wavelengths of radiation that our eyes can't see – perhaps X-rays or infrared – and so the colours of the majestic pillars of gas are necessarily an artistic interpretation.

Hold up your average meteorite against one of these cosmic artworks and, visually, there's no comparison. But here's the thing: meteorites don't lie. You can slice one open, analyse its chemistry and extract a dependable astral testimony that no telescope can ever match. Like those pennies in Winston Churchill's fishpond, these are clues to a realm we can otherwise only see through a glass, darkly. When we clutch a meteorite, we hold the history of the solar system in our hands.

In Part One, we learned how to find meteorites. Now, in Part Two, we're going to look at the lessons they can teach us. Just as many coins are imprinted with the likeness of a country's ruler, meteorites too come with contextual, historical information baked into their internal chemistry. We will see how we can extract this information, weigh it alongside what we know from other spheres of science and thus throw light on several key mysteries. We will look at where Earth got its life-sustaining water and investigate how the impacts of extraterrestrial rocks might have influenced the evolution of life. A crucial part of this involves working out where individual meteorites originated; in other words, matching up the various meteorite groups on the family tree to their parent asteroids. In our last chapter, we will examine the increasingly audacious efforts to make those connections, by sending spacecraft to burgle samples from asteroids and bring them home.

But to begin, let's concentrate on the structure of the solar system itself. It is a rather neat and tidy sort of place, if you stop to think about it. Four small, rocky planets orbiting close to the sun; four large, gassy ones further away. For a long time, this orderliness made perfect sense. When the planets formed, the area closest to the sun was so hot that only rocky materials could survive, whereas further away space was colder and ices could hang around and form into planets like Saturn. Because of this unavoidable temperature gradient, it seemed like any system of planets would naturally end up looking approximately like our own. But the more we inspect the galaxy around us, the more it seems that, instead of our cosmic backyard being natural and normal, it is actually a freak.

Our solar system is also special because it is the only place we know of where living things have emerged and flourished. That makes it all the more crucial to understand how and why it came to have its particular structure, something that scientists have been trying to do for decades. It is a tale that involves fossils of ancient magnetic fields, pinballing planets, leisurely lunches in the Mediterranean sun – and meteorites.

At the grandest scales, the solar system may have a certain logic to its structure, but zoom in a little and you find a hodgepodge of strange worlds and exotica. Even the familiar planets are weirder than we sometimes realise. Closer to the sun than Earth lie Mercury and Venus. Mercury rotates so slowly that one of its days – equivalent to 176 Earth days – lasts longer than its year and Venus has such a thick, cloudy atmosphere that the pressure at its surface is like being 900 metres underwater on Earth. Mars is further from the sun than Earth – it would take about nine months to fly there in a spacecraft. We sometimes think of it as a sister planet to our own, minus the water and atmosphere, but Mars is a planetary minnow, about ten times less massive than our home world.

Beyond Mars and the sprawling asteroid belt we find the giant planets. Jupiter really is huge: it has more than two and a half times as much mass as all the other planets put together. This colossus has upwards of eighty moons (no one is quite sure how many). Those include Io, which has over 400 active volcanoes, and Europa, which has an oxygen atmosphere and an icy crust, thought to conceal an ocean of liquid water. Jupiter also has two groups of asteroids called Trojans that share its orbit, one speeding ahead of the planet and the other behind, like the police bikes in a motorcade. Beyond Jupiter is Saturn, golden and majestic, with its ethereal rings and its own diverse collection of moons. One, Enceladus, has another sub-surface ocean covered with ice, but this time there are also jets of water spewing to heights of 400 kilometres. In 2015, scientists even flew a spacecraft called Cassini right through these space geysers and measured their chemistry.

Then come the real weirdos, Uranus and Neptune. We know barely anything about these two slightly less gigantic giants – having only ever flown one mission past them – and everything we do know is baffling. For one thing, Uranus appears to orbit on its side. Earth's spin axis is tilted at roughly 23.5 degrees from the upright as it orbits around the sun – enough to cause the seasons, but still close to straight. Uranus, by contrast, is tilted by ninety-eight degrees. Imagine the planets as spinning tops. Earth would be leaning slightly; Uranus would have fallen over completely. Both Neptune and Uranus have magnetic fields, but, bizarrely, they are not rooted at the centres of the planets. Earth's magnetic field originates from the swirling liquid iron at its core but Neptune and Uranus might be generating theirs by a completely different mechanism. It might have something to do with rotating layers of electrically conducting ices in their interior – but we simply don't know.

Further out still lies Pluto, now defined as a 'dwarf planet' rather than a planet proper. Part of the reason astronomers decided to downgrade it from planet status in 2006 is that it is part of the Kuiper belt, a vast doughnut of space that contains a lot of other icy bodies, hundreds of thousands of which are more than 100 kilometres wide. The trouble is, if you consider Pluto a planet, where do you draw the line?

The most distant region of the solar system is the Oort cloud. At least, that's what we think. Most comets have such extremely long orbits that we believe they originate in this vast ring of icy debris. No one has directly verified that the Oort cloud exists, but in theory its closest edge is about twice as far away as the Kuiper belt and it may extend a quarter of the way to our nearest neighbouring star. The edge of the solar system is demarcated by the heliopause, the point where the solar wind finally runs out of puff, overcome by the feathery pressure of deep space.

We have a basic understanding of how all this first came to be. In the beginning, there was a vast and formless cloud of dust and gas called a nebula. We think a nearby star then died in a cataclysmic supernova explosion and the shock-waves prompted the nebula to begin collapsing in on itself. In the convulsions that followed, one spot in the nebula ended up with a higher than average density and gravity and then pulled more and more matter towards that point in a snowball effect. Eventually, nearly all the available matter was sucked into a great dense ball: the sun. The remaining material began to swirl around it, flattening into a disc, like a chef whirling dough into a pizza base. This circle of dust and gas – which we call the protoplanetary disc – would eventually glom together into the solar system's planets and moons.

The swirling motion created immense friction, heating the disc to roasting temperatures. Initially, it was so hot that all the matter was present as a gas. However, as the pirouetting gasses cooled, some condensed into solids. Inside meteorites, we find tiny white specks called calcium–aluminium-rich inclusions, or CAIs, which we can date as 4.56 billion years old. We think these were the first solids to form in the solar system, coming shortly before the chondrules, the similarly ancient flecks found inside chondrite meteorites. The formation age of the CAIs is used to define year zero of the solar system.

The high temperatures in the disc also provide the traditional explanation for the solar system's orderly line-up of planets. Space is normally deathly cold, but in the solar system's boiling early days, only rocky materials with a high melting point could survive as solids close to the sun. Compounds like water, methane and carbon dioxide would have been present as solid ices, but only much further away from the sun, where the temperature was lower and the disc was thinner. All this means that the building materials for planets would have been rocky closer to the sun, and icy, gassy compounds further away.

We can rationalise the solar system's orderliness, then. But is it typical? There are approximately 100 billion stars in the Milky Way, and many have their own planets. The first hints that those other systems don't necessarily follow the pattern of our own came in the 1990s, as astronomers got their first evidence of exoplanets; worlds that orbit other stars. You might think of stars as a static point around which everything else moves, but even stars have a small orbit, around the planetary system's centre of mass. This means that, when seen from afar, stars wobble ever so slightly, and the exact degree of wobbliness is determined by the number, size and arrangement of planets around them. Astronomers such as Didier Queloz and Michel Mayor measured these stellar wobbles to glean the first information about exoplanets. These turned out to be like nothing we had seen before: they were about the size of Jupiter, but orbiting so close to their stars that a year lasted just a few days. Since then, we have been able to confirm the existence of these 'hot Jupiters' using other techniques, such as watching the planets blot out the light from their parent stars as they transit in front of them. Today we are even beginning to take grainy images of exoplanets using NASA's James Webb Space Telescope.

If the hot Jupiters sound surprising, they are. A Jupiter-sized planet simply could not have formed that close to its star. At this proximity, the protoplanetary disc would have been blasted by the barrage of radiation emanating from the young star, leaving not enough material to form a huge planet. It would be a while before we understood what was going on here. But from the start, other planetary systems looked radically unfamiliar.

Another piece of evidence making the same point has emerged more recently from a special type of telescope. Point an optical telescope at the sky and you will see things that shine with visible light, not least the stars. That's wonderful, but limited. There are a host of objects out there that don't give off light, but do produce other kinds of radiation, from X-rays to radio waves. One of the most interesting kinds of emission to look at is infrared radiation, the heat coming from objects in space. It is not easy to get a good view of this because the water vapour in Earth's atmosphere absorbs much of the radiation. That is why our best infrared telescopes are often placed either in space or in extremely high and dry locations. This includes the Atacama Large Millimeter Array (ALMA), a set of sixty-six telescopes built about 5,000 metres above sea level in Chile.

When the ALMA began operations about a decade ago, one of its first moves was to look at young stars that still had a protoplanetary disc around them. These give off plenty of heat thanks to all that friction. The first such image, published in 2014, showed the disc around a star called HL Tauri. It was spectacular, an orange circle of dust that had an elegant structure of concentric circles. Astronomers interpret the gaps in the rings as the orbits of still-forming planets, hoovering up the dust as they go. Dozens more images of protoplanetary discs around other stars followed, which turned out to come in a wide variety of patterns – some even shaped like a spiral. 'If you look at this gallery of these ring shapes, no two are the same,' says Steven Mojzsis at the Research Centre for Astronomy and Earth Sciences in Budapest. For him, that provides a clue that tiny differences in the distribution of gas around a star can be amplified to create many diverse kinds of discs – and presumably an array of planetary line-ups.

Back in the late 1990s, scientists were reeling from the sight of those seemingly impossible hot Jupiters, but these weren't the only inexplicable planets. A protoplanetary disc is expected to thin out gradually towards its edges, which implies you should not get particularly large planets forming very far from the parent star. That had astronomers scratching their heads over our very own Uranus and Neptune, which, though smaller than Jupiter and Saturn, are still vast. Orbiting a long way from the sun, at about twenty and thirty times the Earth–sun distance, respectively, these planets are far larger than they had any right to be. Put that together with the hot Jupiters, and it suggested a wild possibility. What if these planets had not been born where we currently see them – what if they had moved? 'At the time, everyone understood that the old ideas didn't work and we needed new ones,' says Hal Levison, now based at the Southwest Research Institute in Boulder, Colorado.

Levison had planet migrations on his mind as he spent a year on sabbatical at the Côte d'Azur Observatory in Nice, France. It is a beautiful place, set high upon a mountain overlooking the city and the aquamarine sea beyond. The main observatory's elegant white dome was designed by none other than Gustave Eiffel, the man behind the eponymous Parisian tower. Levison enjoyed his time there so much that for several years he would return for a month each summer. He struck up a productive collaboration with resident researcher Alessandro Morbidelli. They would enjoy leisurely lunches in the observatory's cafeteria overlooking the sea, discussing the movements of the planets, and then retreat to their computers to run simulations of the history of the solar system. In 2005, Levison and Morbidelli, together with two other scientists, Rodney Gomes and Kleomenis Tsiganis, cooked up a hypothesis that came to be known as the Nice model.

In essence, the Nice model says that all four of the gas giants initially formed in a narrower band of the disc, but then, due to gravitational interactions with each other and with asteroids, they shifted into their current orbits. The model made a big splash because it appeared to make sense of so many riddles. Levison maintains that, to begin with, their motivation was purely to explain how Uranus and Neptune formed – and the model certainly did this. But it also served to resolve other puzzles, such as the existence and structure of the Kuiper belt and why Jupiter has those unusual Trojan asteroids. As Neptune and Uranus travelled outwards, they would have breached through a region of asteroids and rubble, causing complete havoc. 'The whole disc goes ker-plewy,' says Levison. Some of this rocky material was ejected from the solar system, some was forced outwards to form the Kuiper belt and the rest was scattered inwards. As Jupiter then slid towards the sun, it picked up some of that inward-scattered material – and those asteroids are what we now call the Jupiter Trojans. It was this unexpected ability of the model to explain the Trojans, in particular, says Levison, that convinced him the Nice model must be on to something.

The model won plaudits from space scientists, but it was not the final word on planetary migration. In 2009, astronomer Kevin Walsh spent some time studying in Nice with Morbidelli. Naturally, he became interested in his ideas about planets hopping around, but he wondered whether it could have all happened far earlier, perhaps even before the planets had finished sucking up the protoplanetary disc and were still only partially grown. Walsh reasoned that, at this point, the fledgling planets could have been pushed around relatively easily by the gyrating disc of dust and gas – like paper boats bobbing on a pond. Walsh, who is now also based at the Southwest Research Institute, played with what might have happened and in 2011 published a model he called the Grand Tack.

In this telling, Jupiter formed very early and a bit further out than its current orbit, then swooped inwards to roughly where Mars is now. As it did so, Saturn was growing on the outskirts of the solar system and, when it reached a critical mass, its gravity began to suck Jupiter backwards and into its current orbit. The planet had changed direction; or 'tacked' to use the nautical term. All this occurred in just 500,000 years – the blink of an eye, in cosmic terms – well before any of the migrations the Nice model envisaged. The appeal of the Grand Tack model is that it explains why Mars is so small. Working from first principles, you would expect planets forming in Mars's region to be Earth-sized or larger. And indeed, that is what we see when we look at other planetary systems. Apart from hot Jupiters, one of the other kinds of exoplanet we often detect these days is the 'super-Earth', a rocky planet with between two and ten times Earth's mass. According to the Grand Tack, the reason Mars bucks the trend is that the material that would have fed its growth was all gobbled up by Jupiter, during its sojourn in the inner solar system.

The Nice and Grand Tack models both paint the early solar system as a chaotic place where planets behave as if inside some cosmic pinball machine. But does the evidence support this picture? This is where meteorites enter the story. It turns out they have an awful lot to say about the history of the solar system if we study their *isotopes*. It is worth dwelling on these for a moment because we will be hearing a lot more about them.

The basic component of matter is the atom. These contain a certain number of subatomic particles called protons and neutrons clustered in a nucleus, which is surrounded by a cloud of electrons. The number of protons determines which element you have: one for hydrogen, six for carbon, twenty-six for iron and so on. But atoms of the same element can have slightly different numbers of neutrons and these versions of an element are called isotopes. Taking carbon as an example, the isotopes will always have six protons, but there are naturally occurring isotopes that have either six, seven or eight neutrons. These isotopes are named according to the sum of their protons and neutrons. So the most common form of carbon – six protons, six neutrons – is called carbon-12. Various natural processes mean atoms gain or lose neutrons over time, which is why archaeologists use isotopes to date ancient bones or artefacts. But when it comes to extremely ancient materials, like meteorites, their balance of isotopes can reveal the environments they have been exposed to through deep time.

In 2011, Paul Warren at the University of California, Los Angeles, looked at a raft of data on the many isotopes in a wide variety of meteorites. He noticed a pattern: it seemed that all the meteorites could be assigned to one of two groups based on the balance of their isotopes, the most diagnostic ones being titanium, chromium and oxygen. This was the first sniff of a very important result, but it was also arcane stuff. You had to be a serious expert on isotopes to see the logic of Warren's results and even then you had to squint.

Perhaps for that reason, his work flew under the radar for a while. But it caught the attention of Thorsten Kleine, now director of the Max Planck Institute for Solar System Research in Göttingen, Germany. Working with his colleague Gerrit Budde and others, Kleine investigated further, measuring the abundances of other isotopes in meteorites. Looking specifically at molybdenum, he found the same pattern Warren had identified: meteorites sat in one of two groups based on their molybdenum isotopes. In practice, this means that if you measure the molybdenum isotopes in all known meteorites and plot the values on a graph, you will get two lines. Any given meteorite will fit one or the other of these two lines. This time, the split was more obvious than the pattern Warren had noticed and it depended only on measuring one isotope. Molybdenum is present in just about all meteorites too, and as Kleine measured more and more samples it became clear that the whole lot followed the rule. They would either be part of a group of non-carbonaceous (NC) meteorites (which are stony and contain little carbon or moisture) or carbonaceous (CC) meteorites (wet and brimming with carbon-based molecules). To be clear, this division doesn't feature in the decades-old meteorite family tree that we have encountered before. It was a fresh revelation that became known as 'the great dichotomy'.

This dichotomy indicated that all meteorites emerged from two separate reservoirs of dust and gas. As scientists mulled it over and dated the meteorites, it became clear that both groups had formed during the same period, by the time the universe was between one and four million years old. It wasn't that these reservoirs were around at separate times, then. They must have been spatially separated: one region closer to the sun, the other further away. Think of these reservoirs as being like a solar system-sized jammy dodger: you have one reservoir (the jam) in the middle and a second (the biscuit) around the outer edge.

We now suspect the original nebula that went on to form the solar system was built from several supernova explosions, and these different sources account for the isotopic variations. Something kept those two reservoirs segregated for as long as it took the parent bodies of all meteorites to form. What was that something? 'The only likely explanation we could come up with at the time was a large planet,' says Thomas Kruijer, who previously worked with Kleine and is now based at the Lawrence Livermore National Laboratory in California. 'And Jupiter seemed like the best candidate.'

Until now, meteorite analysis and solar system modelling had remained discrete spheres of science, but here, finally, was a chance for them to rub shoulders. If Jupiter kept those two reservoirs apart, it must have been around from very early on to stop the dust mixing. In 2017, Kruijer wrote a paper explicitly making this argument and using the great dichotomy to date the formation of Jupiter. According to his calculations, the planet must have grown to at least twenty times the mass of Earth – not its full size, but still hefty – within one million years of the birth of the solar system. It was the first time the formation of Jupiter had ever been empirically dated and it supported the Grand Tack model, with its vision of Jupiter forming and moving around in the first flush of the solar system's youth.

It didn't end there. The most profound consequence of the dichotomy, says Kruijer, is that it changed how we think about where asteroids – and hence meteorites – ultimately come from. Today, the majority of asteroids orbit in a belt between Mars and Jupiter. The dichotomy showed us that many of them could not have formed there, but instead must have accumulated farther out into the colder reaches of the solar system. That means something must have pushed them inwards – which again, fitted perfectly with the theory of planetary migration.

Recently, meteorites have given us even more insights, this time not into the birth of Jupiter but its adolescence. There are two basic ideas for how this mighty planet could have formed. Option one is a hypothesis called gravitational instability, which says that a patch of gas in the protoplanetary disc collapsed to create the core of the gas giant, which then grew to its full size at lightning speed, within 1,000 years. Option two is a hypothesis called core accretion. In this scenario, Jupiter first forms a solid, rocky, icy core that grows slowly at first until (after perhaps one million years) it reaches a critical mass, at which point it starts to suck in the gas surrounding it at an exponentially quickening pace, ending up as a gassy giant with a solid core. Most theorists strongly favour the second option, but there has never been any direct experimental evidence either way.

Benjamin Weiss at the Massachusetts Institute of Technology in Boston only ended up involved in these matters in a roundabout way. He had spent his career making careful measurements of the magnetic fields in meteorites. It works like this – inside every meteorite, as in all matter, there are electrons gluing the atoms together. The electrons have a quantum mechanical property called spin, which you can think of as a minuscule compass needle that can point up or down. These little needles respond to magnetic fields. If a strong field is present, they will all flip into the same orientation; if a slightly weaker one is around, they do likewise, but alignment won't be so perfect. Even when the magnetic field vanishes, the orientation of these spins remains frozen until some other magnetic field comes along to impart its own stamp on them. This means that meteorites, floating cold and still since the early days of the solar system, can potentially tell us about ancient magnetic fields. 'They are like fossils or echoes of magnetic fields long after they have gone,' says Weiss.

Scientists have long suspected that the protoplanetary disc itself had a respectable magnetic field. Some of the dust was electrically charged and, overall, the disc was moving and churning. We know that these conditions – charged material moving in circles – generate a magnetic field. The same process operates in Earth's core. Many planetary scientists think that gravity alone may not have been enough to drive the formation of the sun and these magnetic fields are needed to explain why our star exists at all. For this reason, Weiss was keen to see if he could provide the first hard evidence of the disc's field by finding its imprint locked inside meteorites.

He decided to try and measure it by looking at chondrules, those first solids of the solar system found as tiny specks inside chondrite meteorites. Doing the measurement was tough for two reasons. First, the magnetic imprint would be extremely faint. Chondrites are generally not the sort of thing meteorite hunters can find with a metal detector. 'A chondrite is not going to stick to your fridge,' says Weiss. Second, the magnetic fields of each chondrule would be pointing in all different directions, so it was no good measuring a whole chondrite meteorite; you would only get a messy average. It was going to be crucial to look at single chondrules, which are typically only a fraction of a millimetre wide. Magnetometers with the necessary spatial resolution and fine sensitivity required had only become available within the past few years. When Weiss and his team turned them onto chondrules, they found exactly what they were hoping for. The echoes of the nebula's magnetic field were there. They measured its strength and found it was about the same as Earth's magnetic field today. If you could go back in time and float around at the birth of the solar system holding a compass, the solar nebula's magnetic field would have tugged on it.

This supports the idea that magnetic fields played a role in forming the sun. But that wasn't all. Next, Weiss and his team analysed meteorites formed at different times to see whether the nebula's magnetic field changed. 'We found that the field vanished after four million years,' he says, concluding that the nebula had dissipated by that time. And here's where we come back to Jupiter. Kruijer had already used meteorites to show that Jupiter grew to about twenty times the mass of Earth within one million years. Now, Weiss's work showed that, from that rocky stage, the planet must have grown to its current mass very quickly. 'I showed that it went from a few tens of Earth masses to about three hundred basically instantaneously,' says Weiss. It was a matter of logic: a growing Jupiter needs the nebula to be present, because that is where it draws its gas from. By the time the nebula had gone, Jupiter must have become the heavyweight it is now.

Meteorites, then, have taught us heaps about the history of the solar system. Jupiter formed very early on and was almost certainly the first planet to do so. It also probably formed according to the core accretion model, as theorists had suspected but never proven. This picture also supports the idea of migrating planets generally, and the Grand Tack model specifically. At the very least, we know that Jupiter does seem to have been around early enough to act as the model suggests. Still, not every scientist is persuaded. Mojzsis, for one, says that the Grand Tack model 'cannot be correct', based on his sprawling analysis of the chemistry of the solar system. For all our progress, the history of the solar system remains stubbornly hard to pin down.

The burgeoning science of meteorite magnetism has more surprises, though. Weiss is currently looking at CAIs, the solar system's first solids, to see if he can deduce when the protoplanetary disc's magnetic field switched on. He has also found younger meteorites that have a magnetic field, but which formed well after the nebula dissipated. These, he says, must have been magnetised by the liquid iron dynamos in the cores of planetesimals. Work like this suggests that, with time and more measurements, we may be able to reconstruct a magnetic timeline of the solar system.

In 2023, Weiss and his colleagues turned to Mars meteorites. Today, Mars has no magnetic field to speak of. The assumption is that the planet, being smaller than Earth, has radiated away so much of its internal heat that its core is no longer liquid and so can't generate magnetism. But we think the core was liquid aeons ago and Mars did, back then, have a magnetic field. If so, that would have protected the surface from harmful radiation and enhanced the chances of life forming on the planet. Studying the magnetism of Mars meteorites could pull back the curtain on all this speculation.

Weiss and his colleagues acquired nine pieces of Black Beauty, the Mars meteorite found in the Sahara in 2011, which we briefly encountered in Chapter Two. It contained crystals that formed on Mars 4.4 billion years ago, which is when we think Mars most probably had a magnetic field. They measured the pieces, but they were in for a major disappointment. The vestigial magnetism of every single one had been erased and over-written by a stronger field. Weiss could tell that they must have been exposed to a strong magnet after they had been found on Earth. Using magnets is a common tactic among hunters to find meteorites, so this wasn't entirely surprising, but it was deflating. These rocks had held a subtle record of Mars's magnetism for as long as the Earth has existed – and then it was wiped in a moment. Weiss says he is now on a mission to persuade meteorite hunters to ditch their magnets, because they are unwittingly deleting the frozen echoes of ancient magnetic fields. There are alternative instruments, such as magnetic susceptibility meters, that could be used instead without doing any damage.

So far, the evidence from meteorites pointed to the truth of the Grand Tack model. But for those paying close attention there was a chink in the armour. The great dichotomy would have kept the two classes of asteroids separated for a long stretch, but then the movements of Jupiter would have shaken and stirred everything, sending some of the outer reservoir spinning inwards and vice versa. It is hard to believe that asteroids from each reservoir never smashed into each other in the ensuing chaos. But if they had, why did all meteorites fit exclusively into one or other of the groups of the great dichotomy? I can't quite believe I am returning to this tortured metaphor again, but if the separate reservoirs were the jam and biscuit of a jammy dodger, why had we never seen any meteorites that contained a mixture of both jam and biscuit mushed together?

A few years ago, Fridolin Spitzer was studying for a degree at Münster University, where Thorsten Kleine, the discoverer of the great dichotomy, was working at the time. Spitzer took on a small research project with Kleine as part of his course. The goal was to look at twenty-six iron meteorites that didn't fit into any existing group on the family tree and measure their molybdenum isotopes. One of the meteorites he chose was a lump of iron that had fallen in 1870 in Nedagolla, India. From the start, it was clearly an oddity. It had some of the highest concentrations of chromium and the lowest of germanium of any space rock. 'It is one of the weirdest meteorites,' says Spitzer. But he wasn't expecting to discover anything particularly telling – it was routine analysis.

Spitzer sat in the laboratory with Gerrit Budde, Kleine's colleague who was supervising him in the lab. They ran samples of meteorites through a mass spectrometer and they would unfailingly plot on either the NC or the CC line, but when they measured the Nedagolla meteorite it didn't plot on either – it was right in the middle. Spitzer remembers Budde saying 'No, this can't be right.' They repeated the tests, but there was no mistake. Spitzer's first instinct was to apologise. 'Budde was the guy who discovered the dichotomy and here was I coming along to spoil it,' he says. But it quickly became clear that this was a sensational result.

The best explanation for the Nedagolla rock is that it formed in a collision between two asteroids, one from the NC reservoir and the other from the CC reservoir. 'Basically, Nedagolla is direct evidence of collision-generated mixing,' says Spitzer. Analysis of the meteorite suggests the impact occurred about seven million years after the birth of the solar system, which broadly fits with the Grand Tack model. Nedagolla is the first and only rock of its kind. I find it a little surprising that we don't have more stones like it, and I put that to Spitzer. He pointed out it might be partly down to the inherent bias we have in meteorites: we only get the ones that happen to cross Earth's orbit. If we do find more like Nedagolla, we may be able to put better constraints on the Grand Tack.

Another test of the Nice model is on the horizon too. Levison is leading a NASA space mission called Lucy that will visit four of the Trojan asteroids in Jupiter's orbit. The mission tips its hat to 'Lucy', the nickname given to the fossilised remains of one of the oldest known human ancestors. The Trojans are fossils too, in a sense, being some of the earliest building blocks of the solar system. Because of their strange orbits, the Trojans are devilishly tricky to get to and the *Lucy* spacecraft will have to take a long and roundabout route, finally arriving towards the end of the 2020s. We have never visited these strange micro-worlds before, so the first pictures will be thrilling. Besides snapping pictures of the asteroids, the plan is to take detailed measurements of their geology. Levison and his team should then be able to discern where in the solar system the Trojans originated. The Nice model predicts that they formed in the outer solar system, but Lucy should finally tell us – two decades after Levison's original papers – if that is correct.

The consensus in the planetary science community is that the giant planets must have migrated. It is possible that the movements required by the Grand Tack happened first and those implied by the Nice model later. The two are not mutually exclusive. And when you put all the ideas and evidence together, you end up with a picture of a solar system that is not neat and tranquil at all, but chaotic and messy. What does that mean for the chances of finding life elsewhere? There is no simple answer to that question. There are a dizzying number of factors that determine whether planets can support life, including how far they are from their star, whether they have a magnetic field, if they have an atmosphere and what gasses they contain. (And that's before we mention water, the presence of which meteorites can, again, shed a lot of light on, as we will see in the next chapter.) But we can say that planetary systems can and do come in a wide variety of structures. Ours seems to be the way it is partly thanks to random chance – and that suggests there could be other star systems out there that, by whatever quirk of fate, have the right conditions to foster living things. But the truth is that, for all our progress, we still understand these matters only dimly.

That much was emphasised by astronomer David Nesvorný, also at the Southwest Research Institute, about a decade ago when he started to look at the Nice model in detail. At the time, the model made many successful predictions, but Nesvorný still felt it needed more rigorous testing. When he made comprehensive computer models to test it, he found they often failed to reproduce key aspects of how the solar system really looks. For instance, the planetary migrations tended to knock away all of Uranus's moons. Nesvorný's simulations would also frequently result in the terrestrial planets being destroyed in the chaos, which is quite the wrinkle. If that had happened, you would not be reading this. 'The first hundred simulations I tried just did not work,' says Nesvorný.

He wanted to see if there was some small tweak that would fix things, so he started tinkering with the simulations, changing some of the starting conditions. One day he wondered what would happen if our solar system had started off not with four giant planets, but five. It was a wild idea and not the sort of thing a self-respecting scientist would propose lightly. 'Inventing planets is not a good thing,' says Nesvorný. 'But this was still a good model that explained many things.' So, he gave it a try. He found that if he placed an additional ice giant between Saturn and Uranus, the model worked like a dream. The planet he had invented was ejected from the solar system completely during the planetary pinball, so it also fitted the empirical evidence. Strange as it sounds, Nesvorný and Levison agree that this five-planet Nice model remains the best explanation for the structure of the solar system.

If it is true, that means one of the founding planets of our solar system is out there, somewhere, drifting in interstellar space, unbound from any star. In fact, Nesvorný believes this sort of thing is common. Estimates suggest there are more of these orphan planets in the Milky Way than there are stars. Nesvorný even gets emails from people suggesting names for the planet that got away. He says he mused on what it should be called now and again, until one day he was watching a movie of *The Cat in the Hat* by Dr Seuss, which features two mischievous characters called Thing 1 and Thing 2. 'From then on, I started calling it Thing 1,' says Nesvorný. That way, he jokes, if we ever conclude that there must have been yet another giant that escaped the solar system, we will have a name ready and waiting.

CHAPTER SEVEN

The Origins of Our Oceans

IN JULY 1966, a Scottish man named Angus Barbieri sat down to a breakfast of a slice of bread and a boiled egg, the first food he had eaten in 382 days. More than a year earlier, he had presented himself to doctors at a hospital in Dundee. He was obese, weighing 207 kilograms, and sick of it, and so he told the physicians he wanted to slim down by fasting. The doctors thought this would be a temporary measure and were happy to advise and observe him. But Barbieri went on and on. He gulped down doses of a few essentials, such as vitamin C, and drank tea, coffee and squash, but nothing else passed his lips. By the time he sat down to his boiled egg, he had lost more than half his bodyweight and two other people could fit into his old clothes alongside him.

Fasting in this way can be dangerous, but Barbieri's story illustrates how surprisingly long a human body can cope without food. Water, however, is a different matter. Stop taking in any fluid and you will be dead in little more than two or three days. Scientists sometimes refer to water as the 'universal solvent', because it is capable of dissolving so many different compounds. That is why it is so essential to life. Most of the essential ingredients of living things – from the carbohydrates that serve as our fuel to the hormones that convey messages around our bodies – will happily mix themselves in among water molecules. So dissolved, they can carry out the complex molecular ceilidh dance that produces self-replicating chemical systems, cells, tissues and ultimately the joyous explosion of creatures and plants on this planet. Water is an absolute prerequisite for life as we recognise it.

It is no coincidence, then, that Earth is both the only place we know of where life has evolved and also a lush world of streams, rivers, lakes and oceans. If you could somehow scoop all the water on Earth into a gigantic sphere, it would have a diameter of around 1,380 kilometres, stretching roughly the distance between London and Rome. Yet, for decades, scientists have profoundly struggled to understand where Earth's water came from. The simplest story of how Earth formed suggests it should be bone dry and the fact that it isn't suggests there is something muddled or missing in our thinking about how planets form and develop. Perhaps there is some undiscovered mechanism that can make a dry planet sopping wet. Straighten this out, and it will be more than an answer to a fundamental question about the world – it should also tell us a lot about which other kinds of planets out there in the galaxy will be damp and therefore capable of harbouring life.

Spend even a little time reflecting on how Earth formed, and you can quickly see why its current wetness requires explanation. Think back again to how rocky planets form from the protoplanetary disc. At the point in the disc where the Earth formed, there shouldn't have been any solid water ice because it was simply too warm and therefore our planet's building blocks, you would think, should have been anhydrous. Even if that were not true, the young planet Earth was a hostile place for moisture. It was blisteringly hot, and had no atmosphere, so wouldn't any water have evaporated away into space? Then there is the not-so-minor matter of the moon. We know that soon after Earth was formed it was thwacked by an object roughly the size of Mars called Theia. Imagine the most cataclysmic space explosion you possibly can, then multiply that by one thousand and you might be close. This collision is thought to have completely vaporised Theia and much of the Earth's outer crust, with a portion of the orbiting debris eventually coalescing into our moon. A cataclysm of this sort does not seem compatible with liquids remaining on Earth's surface.

Simple logic dictates that, if our water wasn't here from the start, then it must have been delivered later. But where did it come from? For decades, the way scientists have tried to answer this question has had our old friend isotope analysis at its heart. The chemical formula of water is H_2O: two hydrogen atoms bonded to a central oxygen atom. But you can also have 'heavy water' (D_2O), where both the hydrogen atoms are replaced by their heavier cousin deuterium, and deuterated water (HDO), which has one hydrogen atom and one deuterium atom. Personally, whenever I think of heavy water I always half-remember a detective story I once read in which it was used as a virtually untraceable way of murdering someone; heavy water looks and tastes much like normal water, but drink too much and it will kill you. For our purposes, though, the point is that scientists have realised that the ratio of hydrogen to deuterium in planets, comets, asteroids and so on varies depending on where they formed in the solar system. Generally, researchers think bodies which formed farther from the sun will have a higher proportion of deuterium than hydrogen.

There is no firm agreement as to why this is so. However, according to modelling studies led by L. Ilsedore Cleeves at the University of Virginia, this enrichment of deuterium in the outer protoplanetary disc came mainly from deuterium flowing in from interstellar space and partly from deuterium-producing chemical reactions that operate at very low temperatures. To complicate things further, it is possible – probable even – that those ratios on any given planet or moon have changed over the solar system's life, so that a deuterium–hydrogen (D/H) ratio we measure today may not reflect the original state of things. On Earth, for example, living things preferentially metabolise hydrogen over deuterium, thus removing the former from Earth's bulk composition. More significantly, hydrogen is far lighter than deuterium, meaning it will be lost to space more easily. Both factors could mean that the ratio of deuterium to hydrogen in Earth's water is now higher than it was aeons ago, by some uncertain amount.

With all that complexity and uncertainty admitted, we can at least ask the simple question: how heavy is the water in Earth's oceans? If you take samples of the ocean in different places the D/H ratio varies slightly, but in 1961 geochemist Harmon Craig decided to come up with a standard sample. He looked at a range of samples of ocean water taken by others in the 1950s, took the average of the D/H ratios and came up with a reference standard called the Standard Mean Ocean Water (SMOW). In this, the D/H ratio is a very small number: 0.000156 (often expressed as 1.56×10^{-4} to make it a little easier to deal with).

There is always going to be way less deuterium than hydrogen, wherever you look – that is just how the universe is made up. What matters is how much there is on one body compared to another. We soon discovered that the D/H ratio of Earth's oceans is really high relative to the rest of the solar system – that is, they contain a large amount of deuterium, relatively speaking. For example, in 1998 Johannes Geiss at the International Space Science Institute in Bern, Switzerland, and George Gloeckler at the University of Maryland, looked at the D/H ratio of the sun. They referred to measurements of deuterium and hydrogen in the solar wind – a proxy for the sun – made by several missions, including by the Ulysses probe which had orbited the sun in the mid-90s, and measurements taken by Apollo astronauts on the moon. This gave them a value of 2.1×10^{-5} for the solar wind, about a seventh of the value for Earth's oceans.

You might have noticed something seemingly awry here. Earth's oceans are made of water but the sun and the solar wind distinctly aren't. The comparison isn't quite as bananas as it might sound, though, because of the rock cycle. Throughout Earth's history, rocks from the mantle, the hot, gooey layer beneath the crust, are brought to the surface through volcanic eruptions. There they are weathered and hydrogen and oxygen atoms can equilibrate between being joined together in water molecules and incorporated into the structures of rocks. Eventually, those rocks can get washed into the seas, compacted by the pressure of the ocean and squeezed by the tectonic plates, and become part of Earth's innards again. In this way, there is a subtle but undeniable connection between lone hydrogen (and deuterium) atoms and water molecules and so we use hydrogen measurements as a proxy for water. It isn't perfect, but it's a start.

Let's return to the central question of how Earth got its water. It could be that once the moon-forming cataclysm was over and the planet had cooled a bit something delivered the moisture. This general idea is known as the 'late veneer' because, even though it delivered a vast amount of water and other volatile chemicals, it was just a thin layer of material in the grand scheme of things – a coat of varnish on an almost-finished planet. One delivery vehicle scientists have eyed for decades is comets. At first, this was a natural assumption rather than a carefully considered theory. Comets are often called 'dusty snowballs'; they are largely made of ice and they could have pelted the young Earth and delivered the water.

The idea got its first real test in 1986, when Halley's comet passed the Earth. Halley's is perhaps the most famous of all comets. We have known of it for centuries: it appears on the Bayeux tapestry, which depicts the Norman conquest of England in 1066, and the Italian painter Giotto observed it in 1301 and depicted it shooting above the Bethlehem stable where Jesus was born, in his painting *Adoration of the Magi*. It was not until 1705, though, that the English astronomer Edmond Halley realised it was the same comet appearing periodically every seventy-six years – it is from him that the comet takes its name.

Comets were once seen as harbingers of doom and this was the case even as late as 1910, when Halley's comet appeared once more. At the time, scientists took a spectrum of the comet, meaning they measured the wavelengths of light passing through its coma, the diffuse cloud of dust surrounding its solid core. Different molecules absorb light in characteristic ways and the spectra showed that Halley's comet contained cyanogen, a toxic gas closely related to cyanide. The *New York Times* ran a piece that year announcing this, and musing on what might happen if Earth's orbit swung through the comet's wake, with an astronomer named Camille Flammarion as its principal source. 'Prof Flammarion is of the opinion,' the reporter wrote, 'that the cyanogen gas would impregnate the atmosphere and possibly snuff out all life on the planet.' Suffice it to say that those fears proved unfounded.

In the early 80s, in the lead-up to Halley's return, scientists wanted to take a closer look. Astronomer Kathrin Altwegg, now at the University of Bern, Switzerland, says that at the time they just wanted to know the basics about comets and what they were like up close. Having never visited one before, there were many questions. For instance, it was unclear whether they even had a solid nucleus or if they were dusty clouds through and through. They also wanted to know what they were made of, including their D/H ratio. So the European Space Agency launched a mission named

Giotto, after the painter, with Altwegg as a young scientist on the team. They took images of the comet demonstrating that it did indeed have a solid nucleus, and using spectroscopy the team found that it was about twice as rich in deuterium as Earth's oceans. Isotopically, comet water seemed to be far removed from that of Earth and scientists wrote the theory off. 'They really thought it meant that our water does not come from comets,' says Altwegg.

Then again, that was just one comet. Over the next few decades, telescopes were trained on other comets as they passed us from time to time and their D/H ratios were measured too. In 1998, for instance, researchers used the James Clerk Maxwell Telescope in Hawaii to measure the spectra of the Hale–Bopp comet. It turned out to have a similar D/H ratio to Halley's comet and, by the 2010s, we had measured a clutch of six cometary D/H ratios, all clustered around roughly the same value. It seemed that comets really couldn't be the source of Earth's water.

Then in 2011, there came a surprise. Planetary scientist Paul Hartogh and his colleagues looked at a different sort of comet. It is thought that there are broadly two kinds of comets. Long-period comets have, as the name suggests, very long orbits that take them far out towards the edges of the solar system. So far, all the comet D/H ratios we had were from this type of comet. But Hartogh was interested in comet Hartley 2, which comes from the so-called Jupiter family, which is thought to be a distinct group with much shorter, tighter orbits. When Hartley 2 passed us in 2010, Hartogh and his colleagues took its spectra using the Herschel Space Observatory. They found its D/H ratio was

1.61×10^{-4}, which, allowing for a margin of error, put it right on top of the SMOW value for Earth's oceans of 1.56×10^{-4}. It suggested that even if the long-period comets couldn't have brought water to us, other kinds might have. Comets were back in the frame as a possible origin of Earth's water.

All this added to the excitement around the Rosetta mission, which had been launched in the early 2000s and was due to land on comet 67P/Churyumov–Gerasimenko, another Jupiter family comet, in 2014. Altwegg was the scientist behind the craft's ROSINA spectrometer, which made the crucial D/H measurement. And sadly it wasn't good news. Comet 67P wasn't just high in deuterium, it had the most of any comet yet measured, about three times as much as Earth's oceans. 'I think a lot of people hoped that comets would have terrestrial-like water,' says cosmochemist Conel Alexander at Carnegie Science, a research institute in Washington DC. 'But this result confounded them.' Altwegg has a similar view. 'After this, comets were really out,' she says. Some people still think they may have contributed a little of Earth's water – perhaps 10 per cent – but nothing serious.

You might be wondering when meteorites are going to feature in this tale. We are getting to that, and to cue it up there's another side of this multifaceted riddle we need to look at. When we ask where Earth's water came from, we are also really asking a whole set of other, related questions to do with other elements. If we believe the argument that tells us Earth should have been born dry, this suggests it should also have been born without other volatile elements, such as helium and nitrogen. Come to that, Earth is sprinkled with particular concentrations of the over 100 other elements in the periodic table, from astatine to molybdenum and niobium to xenon. Were they here from the start or were they delivered later? If we come up with some hypothesis that explains the origins of Earth's water, but predicts the wrong concentrations of the volatile gas xenon, say, then it can't be right.

Within the small group of scientists who study these matters, some are interested in the highly siderophile elements (HSEs). This group, which includes some that we hold most precious, such as gold and platinum, has an interesting property that's been established through experiments. If you were to take a chunk of iron sprinkled with elements and a chunk of rock sprinkled with elements, glom them together and melt them, the HSEs would all partition themselves into the metal. 'Siderophile' means 'iron-loving' and this effect is so strong that more or less all the gold, platinum and other HSEs move into the metal with almost none left in the rock.

This matters because Earth performed this iron and rock melting trick approximately 4.5 billion years ago when that Mars-sized object Theia hit us and created the moon. The impact is assumed to have been so energetic that Earth's insides melted, with a rocky liquid mantle floating on top of the inner liquid iron core. All the HSEs would have then promptly slipped into the core, resetting their concentration in the mantle to zero. 'The moon-forming event would have cleaned out the highly siderophile elements from the mantle,' says Richard Walker, an expert on the siderophile elements at the University of Maryland in Baltimore. Here's the point: when we look at mantle rocks today, any HSEs we find must have been put there *after* the moon formed.

It is possible to get hold of rocks from the mantle thanks to something called mantle plumes. These are pipes of hot rock that rise from the mantle to near the surface in a few locations scattered around the globe. In 1978, geochemist Chen-Lin Chou looked at some of these rocks – which are known as ultramafic rocks – and measured the concentrations of twenty siderophile elements in them. He found that the profile of HSEs in the mantle rocks was a good match for the CI-chondrites, a type of carbonaceous chondrite. All you would need is a gentle puttering rain of these meteorites – equivalent to about one per cent of the weight of Earth's mantle – and this would explain its elemental composition. This observation, says Walker, led to the idea that Earth must have been peppered with carbonaceous meteorites (some so large you'd probably call them asteroids) shortly after the moon formed.

In 2003, the cosmochemist François Robert added water to the mix. He looked at the D/H ratios of the carbonaceous chondrites and showed they were a good match for Earth's ratio. The hypothesis was that these meteorites had the right mixture of elements and water to match what we actually see on Earth. Many scientists still believe this pretty much explains how Earth got its water. 'If you take all the different bits of evidence, the carbonaceous chondrites are probably the best match,' says Alexander.

All settled then? Not so fast. In 2004, Karen Meech, based at the University of Hawaii, was in Iceland for a scientific conference covering her field of astrobiology. Big international science conferences sometimes come with the option for delegates to make an excursion to some tourist spot in the local area; a treat for those who have travelled a long way. Meech decided to make the most of the opportunity and went on a trip to some hot springs in the north of Iceland. There, a tour guide explained how the springs bubbled up from underground and told Meech and her fellow day trippers that this was 'primordial water'.

There and then 'a lightbulb went on over my head', says Meech. She had long been pondering the origins of Earth's water herself and wondering if the D/H ratio of Earth's water might have changed over time. When we looked at the D/H ratio of comets and asteroids, we compared it with the SMOW. But was that comparison legitimate if the relative amount of deuterium on Earth had changed over time? Here, Meech realised, there was potentially a way to find out. Get a sample of 'primordial water' that had been locked away in Earth's depths since the planet's formation, and that might give us a better sense of what its original water was like. 'I thought, oh, we've been measuring in the wrong place,' says Meech. 'And I was suddenly really excited.'

The tour guide had been mistaken, it turned out. Iceland's hot springs are geothermally heated, but the water isn't 'primordial' in any meaningful sense; it's ground water that has percolated upwards. But that didn't stop Meech. She began to think of the rocks brought to the surface on mantle plumes, which we encountered earlier, and asked around to see if any other scientist had ever measured the D/H ratio of such rocks. No one had, colleagues told her, because it's a tough measurement that you can only really do if you have a specialised, expensive instrument called an ion microprobe. At this point Meech's excitement reached fever pitch, she told me, because her department had recently taken delivery of just such a machine.

The stars seemed to be aligned in Meech's favour and she investigated two possible sources of ultramafic rocks. She looked at some from Iceland, which sits on top of a huge mantle plume, and she also enquired with a scientist named Donald Thomas at the University of Hawaii, who had an experiment that involved drilling into the plume that wells up beneath Hawaii. Thomas gave her some rock samples, but it was clear, says Meech, that both these and the rocks from Iceland had come into contact with water on their way up to the surface. They may once have held primordial water, but it had been too contaminated with fresher stuff, making them useless for her purposes.

In 2014, Lydia Hallis, now based at the University of Glasgow, went to work for a while with Meech in Hawaii and took on the project looking for primordial water samples. The first two possibilities Meech had investigated were duds, so Hallis trawled the scientific literature looking for reports of other places she could find rocks from very deep within the Earth. She came across some ultramafic rocks recovered in 1985 in Baffin Island, Canada. These had come to the surface about 60 million years ago on another mantle plume. Hallis wrote to the scientist who had collected them, Don Francis at McGill University, and asked if she might have a small sample. Francis said sure; he was about to retire and he had loads of the rock left over. He sent her a whole box.

These rocks were seriously ancient: dating techniques put them at approximately 4.45 billion years old, so only a few tens of millions of years younger than the age of the Earth. Hallis could also tell that they had hardly been affected by their passage through the crust. And there was more. Minerals or rocks can form with bits of other material locked within them, which geologists call inclusions. Inclusions are special because you can tell they have not been disturbed since at least the time when the mineral itself formed. In the rocks from Baffin Island, Hallis found tiny glass inclusions within chunks of olivine, a mineral that forms in the mantle. These inclusions were so old that they could preserve the signature of Earth's D/H ratio as it was when the planet first formed. Hallis and Meech used the ion microprobe instrument to measure the D/H ratios in these glass inclusions.

The results were not straightforward to interpret. Hallis observed a spread of D/H ratios in these inclusions, which she takes to mean that hydrogen from the mantle had seeped into them to varying degrees as they ascended. This means she doesn't have a pure measurement of Earth's primordial water. However, she could plot her measurements on a graph that accounts for how much mixing each sample has undergone. Based on this, she could extrapolate backwards to estimate the D/H ratio of a pure sample of primordial water. The figure came out as 25 per cent lower than the water in Earth's oceans. The accuracy of that number is uncertain, but it does provide evidence for what Meech and others have long suspected: Earth's D/H ratio changed over time and was originally closer to that of the sun.

What did this mean? Hallis says it indicates that Earth might have got its water by directly accreting hydrogen gas from the protoplanetary disc in its first flush of youth. This would have been incorporated into Earth's inner rocks and would only come to the surface later, through volcanism. Slowly, the ratio of deuterium to hydrogen could have changed as the lighter hydrogen evaporated more easily into space, presenting us with the values we measure in the oceans today.

As I interviewed Hallis about her work, I realised that if her interpretation was true, it shattered a lot of the assumptions I had read about water and Earth's formation. For example, this early-accreted water would have had to survive the moon-forming impact – which seemed crazy to me. But as I spoke to Hallis, I realised that my assumptions were just that. Arguably, there's no reason to think that water could not have remained gravitationally bound to Earth after that cataclysm. Plus, Hallis points out that we know there is water on the moon itself, and that this formed from a cloud of completely vaporised debris. 'My work suggests that there is no need for Earth to form dry,' says Hallis.

Hallis's work was controversial, however. 'There were people who violently and publicly disagreed with me,' she says. She remembers one awkward conference where the speaker before her talk spent his full allotted time explaining why he thought Hallis was misguided. For his part, Alexander says he doesn't entirely buy the idea that the hydrogen could have come from the protoplanetary disc. 'I'm not sure if many people believe that any more,' he says. One issue, he says, is that Hallis couldn't find a pure sample of primordial rock and was forced to use those extrapolations. Another niggle, he says, is that she only looked at rocks from one plume and in principle they could be anomalous. The results would be more persuasive if they were based on rocks from plumes across the planet.

If you are feeling bewildered by all this, you are not the only one. 'It feels as if, over the past few years, every big paper in planetary science that has made it into the most prestigious journals has basically said "this matter of Earth's water, you know what, it's more complicated than we thought",' says Luke Daly at Glasgow University. We first met Daly in Chapter Three, when he was part of the team chasing down the Winchcombe meteorite. But he has also done some work that adds another brush stroke to our portrait of how Earth got its water.

There's always been a loosely held assumption that water couldn't have been around as a solid in the protoplanetary disc when Earth was forming because it was just too hot. Yet what if that's not so? Planetary scientists have long known about a phenomenon called space weathering. Just as rocks on Earth are battered by the wind and rain, rocks in space are assaulted by tiny dust grains, cosmic rays – super-high-energy particles that zip across the universe – and, most significantly, the solar wind. Since the solar wind contains hydrogen ions, scientists have pondered for decades whether that hydrogen could pepper itself into asteroids and combine with the oxygen already present in their rock to produce water. If this is so, water wouldn't have to exist in the protoplanetary disc as motes of ice, but could possibly accumulate on the surfaces of grains of dust.

This amounted to speculation until 2018, when a team led by John Bradley at the University of Hawaii did some simple experiments. They took some powdered and completely dry olivine, a mineral containing iron, silicon and oxygen that is the principal constituent of Earth's mantle. Using this as a proxy for a space rock, they put it in an extremely cold, airless chamber and blasted it with a beam of hydrogen ions, to simulate the solar wind. The team found that this on its own was not enough to produce any water. However, they then repeated the experiment, this time also shooting the olivine with laser pulses, designed to mimic the energy imparted to the rock in space when it is hit by micrometeorites. This time, water was created. The team wrote that this was an 'efficient, versatile pathway to generate and liberate water'. Bradley and his colleagues also looked at interplanetary dust particles (IDPs) and found that these contain water too.

All this suggested that water could form on rocks through space weathering. What remained uncertain was whether in practice this was just the odd drop of moisture or an amount that could potentially contribute significantly to a planet's seas. Enter Daly and his colleagues. In 2021, they took three tiny grains brought back by the Hyabusa sample-and-return mission from the asteroid Itokawa – more of which in the next chapter – and studied them using atom probe tomography. This technique has been used to analyse the properties of man-made materials for decades, but it has only recently been modified so it can be used on rocks.

It's a powerful tool: it works by evaporating individual atoms from the surface of the rock, enabling scientists to generate a map of its surface at the atomic scale. By pulling off more and more atoms, you can also drill down into the rock and see how the composition changes as you go. 'You can see the entire periodic table', says Daly. You can find out where you have water in your material and how it varies by depth.

Daly says that when they analysed grains from the asteroid, they did indeed find they contained water down to a depth of about sixty nanometres from the surface, a distance roughly ten times thinner than a human hair. That fitted with his expectations, since the solar wind would have penetrated only a very short distance into the rock. After making his careful atomic map, Daly found that about 1 per cent of the molecules in that thin surface layer were water. Extrapolate that throughout the whole of Itokawa, and he reckons you're looking at about twenty litres of water per cubic metre. It was strong evidence that dry asteroids can indeed be seriously moistened by the solar wind. Since Itokawa is one of a fairly common kind of asteroid, Daly and his colleagues wrote that 'such water reservoirs are probably ubiquitous on airless worlds throughout our galaxy'.

Now, twenty litres of water per cubic metre of asteroid might sound a lot, but there's a nuance here that takes us further. Daly also looked at extremely tiny grains from Itokawa too, specks of dust no bigger than one micrometre across. In these cases, it seems the solar wind was able to penetrate the whole of the grain, so that the entire speck was enriched in water to about the same degree as the thin outer layer of the grains. To see why that matters, imagine you have two boxes balanced on a scale. Into the first, you put a cubic metre of solid rock from the asteroid Itokawa. Then you fill the second box with tiny grains of dust from the same source so that the scales balance. You have the same mass of material in each box, but the first only has a thin veneer of water-enriched rock, whereas the dust in the second box is enriched through and through. If then you imagine that dust falling on the young Earth, well, it could have delivered a seriously important amount of water.

As we saw when we looked at Jon Larsen's work on micrometeorites, scientists have long guessed that Earth receives a constant rain of cosmic dust. Daly's work suggested this could have been a significant source of Earth's water balance and was worth considering alongside asteroids. And for Hallis, who worked with Daly on some of these studies, there's another implication. The protoplanetary disc would have been full of silicate dust and the same space weathering mechanism could have produced a lot of water on those grains during the birth of the solar system and the planets. It is, she thinks, another reason to take seriously the idea that Earth could have accreted a lot of its water very early on.

There is also a radical school of thought emerging among some who study the origins of Earth's water, one that presents a dramatically different perspective on all the arguments we have heard so far. For a long time, the leading idea for how planets formed proposed that it happened in two phases. First, dust and tiny particles slowly glom themselves together into kilometre-sized chunks called planetesimals, and second, these planetesimals collide with each other and coalesce to form planets. But in 2012, Michiel Lambrechts and Anders Johansen, two planetary scientists at Lund University in Sweden, put forward a new idea called pebble accretion, applying it first to the solid cores of the gas giants but later also to rocky planets. There's nothing too complicated about it. It simply says that pebble-sized objects would have accumulated steadily into planets, rather than necessarily having to form multiple planetesimals that then collide. 'It's a bit like a car accident on a foggy road,' says planetary scientist Zachary Sharp at the University of New Mexico in Albuquerque. 'Two things collide and then more and more keep on piling into the back of them until you have a huge pile-up.'

The point about pebble accretion is that it can build a planet much faster than the traditional theory would predict. Previously, we assumed it took tens of millions of years after the formation of the protoplanetary disc for Earth to grow to an appreciable size. But pebble accretion predicts Earth grew to around 70 per cent of its present mass in two million years or so. At that time, the protoplanetary disc would still have been relatively thick and so Earth's burgeoning gravity would have drawn a cloak of gas around it. This atmosphere would have been rich in hydrogen and would have built up to an incredible pressure. That would have heated the Earth, melting its rocks, whereupon some of the hydrogen gas would have dissolved in them. The ensuing chemical reactions would have created a lot of water, which, in the fullness of time and after being brought back to the surface by volcanic activity, would be the source of Earth's oceans. This, anyway, is how Sharp envisages it. If Earth formed much earlier than we thought, it suddenly makes this whole other narrative possible. 'I think it's revolutionary,' says Sharp. 'It's up there with like plate tectonics or something, because it makes sense, it works and it just explains so much.'

Alexander says that pebble accretion has got a lot of planetary scientists excited and that it may well have played a role in forming the cores of the giant gas planets. However, he is not so sure about applying it to our own planet. If Earth did form as early as Sharp thinks it did, and if it did accrete an atmosphere from the solar nebula, then that would have included not just hydrogen but also smaller amounts of other gasses such as nitrogen and noble gasses such as xenon. Alexander points out that when we compare the isotopic composition of those gasses in our atmosphere to that of the solar nebula, they don't match. 'For the moment, I remain sceptical,' says Alexander.

At this point, I want to level with you; reporting this chapter of the book was tricky. Usually, when I start investigating a story as a journalist, I make a list of a few key questions, interview a handful of experts and build up a sense of what the scientific consensus is. It is never the case in science that everything is unanimously agreed, but often there's a core idea that most people think is roughly right. But the origin of Earth's water isn't like that. The more phone calls I made, the more different points of view I encountered and, truthfully, the more confused I became. Often it felt as if one view was backed up by reams of isotopic evidence from a particular set of elements, while another opposing set relied upon a separate and contradictory tranche of isotopic evidence from another set of elements. Sorting out the truth is like trying to unpick a series of complicated knots. One scientist ended an email to me by writing: 'I realise what a confusing subject this is, but it's mostly because we don't have all the answers yet!' You can say that again.

That is not to say we have made no progress and indeed, I think this is one of those stories where the journey is more important than reaching the final answer. Many scientists said to me that Earth's water probably arrived in lots of different ways: maybe a tiny splash from comets and a sprinkling from meteorites and dust and some of it perhaps was here from the very start. You might remember that my own interest in meteorites was piqued when I went to a conference about the origins of water in the solar system. This was back in 2017, and it was organised by a group of scientists including the Natural History Museum's Sara Russell. The idea was partly to try and thrash out a consensus. In their summary report, the conference organisers wrote: 'a consensus emerged that volatiles were likely incorporated into the terrestrial planets both during planetary accretion and later by asteroidal impacts'. Having spent many months talking to people about this matter, I am not so sure there is as much of a consensus as that sentence implies – but perhaps it is as close as we can get.

But let's zoom out a bit. We have learned that, at the very least, the assumption that rocky planets must be born dry is questionable. Plus, there are a range of ways in which water can be delivered later in their lives. It's not so much that rocky planets are naturally dry and we have to wrack our brains for possible ways in which they could be watered. If anything, it's the other way around: star systems seem to have a plethora of mechanisms at their disposal to douse planets in water. That potentially bodes very well for the chances of finding water – and therefore life – on planets circling other stars. For Hallis, the idea of space weathering is particularly exciting in this respect. 'If we can say there is a significant source of water from space weathering, then that would affect every rocky planet around every other star,' she says. Not only that, but the history of the exploration of the solar system over the past two decades has shown us that liquid water crops up in all sorts of unexpected places, including many of the moons of Jupiter and Saturn. All the evidence encourages us to think that watery worlds should be commonplace throughout the galaxy.

There is a coda to this story. Up until now, we've been dwelling mostly on water – or its proxy, hydrogen – embedded within rocks and using such clues to reconstruct where Earth's water came from. But it turns out that meteorites can sometimes also contain a secret cargo of actual liquid water trapped inside them. This comes in the form of fluid inclusions, which are much like the inclusions that Hallis measured in her mantle rocks, except they are liquid rather than solid. These tiny treasures, which are sometimes called 'flincs' for short, give us a unique window on the watery past of our cosmic neighbourhood.

The year 1998 was an auspicious one for flincs. In March of that year, a meteorite fell in the arid town of Monahans, Texas. A group of boys happened to witness the fall and a piece of the meteorite was recovered straightaway. Mike Zolensky, who works on the other side of the state as curator of meteorites at NASA's Johnson Space Center in Houston, got hold of a piece of the meteorite and had it in one of his clean rooms within forty-eight hours. There, he broke it open with a hammer and chisel and saw inside some crystals of a mineral known to scientists as halite – and to the rest of us as common salt. Any crystals of salt are very rare – almost unknown – in meteorites and these were especially unusual as some were large, up to a centimetre long.

A few months later, in August, another meteorite fell in the desert close to the small town of Zag in the south of Morocco. It was another extremely dry area and, again, the meteorite was recovered rapidly, before any rain had fallen (not that there ever was much rain to speak of in Zag). Meteorite dealer Edwin Thompson – everyone calls him ET – acquired the stone and broke it open. He'd heard about Zolensky's find from a few months before and when he saw that it contained halite grains, he quickly sent the NASA curator a sample.

With both the Monahans and Zag meteorites in his hands, Zolensky looked at them under a microscope and saw that they contained tiny spaces inside the salt crystal where a drop of fluid was trapped. He could tell, because there was a tiny bubble of gas in there too, which moved around – it was almost like looking at a miniature spirit level. These fluids were sealed fully inside the meteorite, meaning that they had been there since the rock formed and the salt crystallised, during the youth of the solar system. Zolensky asked a colleague to date the halite in the Monahans meteorite and it came out as at least 4.5 billion years old. This appeared to be a direct sample of water captured at the very birth of the solar system.

When Zolensky looked into it, he found that another scientist had encountered such things before. In 1864, the English geologist Henry Clifton Sorby, a pioneer of the microscopic study of rocks, had published an account of fluid inclusions in meteorites. Yet it seems almost no one took any notice of this or followed it up. Then in 1983, out of the blue, a group of researchers announced they had discovered fluid inclusions inside no fewer than five meteorites. Sadly, within a few years, their work had been shown to have a fatal flaw. Meteorites are frequently cut into slices with saws so they can be analysed and solvents are used to lubricate the process. Another group of researchers noticed that when they did this tiny bits of the solvent would sometimes get sucked into microscopic pores in the meteorite in a way that made them look a lot like fluid inclusions. This, Zolensky later wrote, would lead to 'reports of fluid inclusions in meteorites being viewed with scepticism for many years'.

Anyway, in Zolensky's view, that scepticism was not valid for the Monahans and Zag stones. Those had been collected before any rain fell and the meteorites had been opened up carefully. He had used only a hammer and chisel and Thompson had used a saw lubricated with pure methanol – no water. 'The possibility that the fluids were introduced into the halite after reaching Earth is nil,' he wrote. The halite crystals were also a bright blue colour, due, Zolensky concluded, to long exposure to cosmic rays in space. This showed the halite crystals were very old and not of terrestrial origin.

The next step was to isotopically analyse the fluid inclusions. Zolensky and his team carefully picked out the tiny halite crystals using a stainless-steel needle, then analysed them using a mass spectrometer, which separates out ions by mass and measures them. The results were odd: the eight flincs they managed to measure, taken from both meteorites, were all vastly different in their D/H ratios. This, Zolensky later wrote, must mean these liquids were in 'disequilibrium', the scientific way of saying that they had not had time to thoroughly mix. This led Zolensky to rather a wild conclusion: that the flincs must have come from a cryovolcano.

Let's unpack that. Regular volcanoes on Earth spurt out liquid rock, but on other, colder worlds there could be an analogous process where water ice is squeezed, heated and erupts. This might sound fanciful at first blush, but in fact there's no 'could be' about it. We know that cryovolcanism happens. In 2015, NASA's *Dawn* spacecraft arrived at Ceres – which I think of as the largest asteroid but which, at 940 kilometres wide, is technically a dwarf planet – and began photographing its surface. The images revealed a four-kilometre-high mountain with bright streaks running down its sides, which scientists named Ahuna Mons. It looked an awful lot like a volcano. Planetary geologist Ottaviano Ruesch and his colleagues analysed and modelled the structure and concluded it must be an ice volcano, which was dormant now but was active as recently as about 210 million years ago. We now think Ceres has more than twenty similar cryovolcanoes. And it isn't the only place that has such activity – Enceladus is also venting icy brines into space, even today. The forces that spur such activity are probably different from those responsible for volcanism on Earth. One possibility is tidal forces, for example, where the gravity of an orbiting body pulling and squeezing rocks could create friction and melt the ice.

Around 200 million years ago on Earth, you could have witnessed the biological wonders of the dinosaurs. If you had stood on the pockmarked surface of Ceres at the same time, you might have witnessed a wonder of the geological variety. It would have started with a rumble. Melted salty brines would be rushing upwards from somewhere beneath your feet, before popping out of the top of Ahuna Mons in an icy blast. It might have happened so quickly that the brines would not have had time to mix. Then, as globs of brine were sprayed into space, they would have rapidly cooled and crystallised. Over time, those drifting clumps of ice must have struck a piece of rock and been melded into it, before falling to Earth as a meteorite. Or at least, this is roughly how Zolensky thinks the flincs in the Zag and Monahans meteorites were created. 'These samples probably aren't from Ceres, but they're from a body just like Ceres,' he says.

What would be wonderful would be to do comparative studies looking at flincs from different materials and formed at different times. To do that, we need more of them. And there are people looking. At Royal Holloway, University of London, Queenie Chan and her PhD student Peter Krizan have been conducting a project to think through what kinds of meteorite might feasibly contain flincs. It's been a matter of carefully narrowing down the options, requesting samples and then making careful studies of specimens. 'It just requires a focused project like mine to look for them,' Krizan told me. 'No one has really spent time looking for them before, but now it's come up that this is possible.'

For his part, Zolensky says he has found another fluid inclusion – in this case, a mixture of carbon dioxide and water – in a meteorite called Sutter's Mill, a carbonaceous chondrite that fell in California in 2012. 'It's just a matter of time before Peter finds more,' he says. 'They're there, they're just really hard to find because they hardly ever survive.'

Because the inclusion in the Sutter's Mill stone contains carbon dioxide, that means it must have formed in a region of the solar system where it was cold enough for this compound to freeze solid – so near where the gas giant planets are or even further out. From there, we know it migrated to the asteroid belt and eventually to Earth. 'So this is additional evidence of the large-scale migration of these icy bodies,' says Zolensky. He's talking about the kinds of movements predicted by the Nice model.

Not only that, but the flincs in the Monahans meteorite are loaded with organic chemicals, which Zolensky finds particularly exciting. 'Think about it,' he says, 'these are fluids at room temperature, they're brines, like in Earth's ocean, and they are loaded with organics like amino acids.

All the building blocks for life are there four and a half billion years ago in some reservoir.'

Personally, just on an aesthetic level, I find the idea of fluid inclusions deeply appealing – water locked away, pure and perfect since the solar system began (or close enough). But researchers I have spoken to have conflicting views on how useful any of Zolensky's discoveries will ultimately prove. Alexander says it is difficult to see what fluid inclusions could really tell us. On the other hand, Hallis finds it thrilling to imagine where this line of research could take us. I asked her what it would be like if we could find flincs from a range of different places. 'That would be fantastic, to have fluid inclusions from all different bodies,' she told me. 'It would open up a whole new world of science.'

CHAPTER EIGHT

Space Fossils

HE IS MOSTLY over it now, but Birger Schmitz once had a strange compulsion. Any time he was in a railway station, airport, bank – any public building with a large expanse of stone floor would do – you would find him on his hands and knees, eyes glued to the floor. 'I have had problems with security guards,' he admits. 'They tend to get suspicious if you start crawling around in the dark corners of an airport.'

Schmitz is no terrorist. He is the world's foremost expert in fossil meteorites. These treasures are unthinkably rare and difficult to find. Getting hold of an ordinary meteorite is hard enough but imagine trying to recover one that fell millions of years ago, before being entombed in solid rock like the bones of a dinosaur. An unlikely sequence of events led Schmitz to realise stone floor tiles were one good place to start. And then, over the years, he realised there were other ways to snare these rarities and went on to find them by the dozen.

This hunt is worth it. While freshly fallen meteorites present us with a potentially unblemished record of the solar system's history, fossil meteorites – if you know how to read them – tell a subtle story about the interaction between Earth and space through the deep past. Most of us know that the dinosaurs were done away with by the impact of a huge asteroid, but Schmitz's efforts, and increasingly those of other fossil meteorite hunters, illuminate how rocks from space may have had a profound impact on the course of Earth's history – not just once, but potentially many times.

To tell Schmitz's story, it helps to start with the asteroid that killed the dinosaurs and the work of a scientist named Luis Alvarez. It was he who, in the late 1970s, found the KT boundary, a layer of ancient rocks in which lay the traces of one of the most epic explosions ever. Eventually the source of this explosion was pinned to a giant asteroid called the Chicxulub impactor, which slammed into the Earth some sixty-six million years ago, doing away with nearly all the dinosaurs, not to mention countless other living things. When dust from the impact settled, it created a new layer of detritus across the planet enriched with elements such as iridium, which are normally rare on Earth, and Alvarez had found this layer. This hypothesis is now largely accepted, but at the time it was brand new.

Schmitz, however, wasn't buying it. In 1985, he published a scientific letter saying that the iridium could just as easily have come from seawater. 'I challenged him,' says Schmitz. It was a bold move for a young researcher, particularly considering that Alvarez was established as an eminent scientist, having won a Nobel prize more than a decade earlier for separate work in physics. Alvarez saw the letter but didn't react badly. Instead, when he had cause to visit

Schmitz's native Sweden in the autumn of 1986, he asked if his cheeky correspondent would be present at one of the dinners he was attending.

In Schmitz's telling, Alvarez was a man who liked to challenge other scientists and be challenged himself. His approach was always to look for ways in which his ideas could be wrong and he appreciated the fact that Schmitz was ready to question him. At the dinner, he asked Schmitz if he would come and be part of his research group at the University of California, Berkeley. Part of his role would be to play devil's advocate, trying to spot potential flaws in Alvarez's work. Schmitz was glad to do it, but before he had been there long the evidence had built up to the point where the extraterrestrial cause of the KT boundary was undeniable. A few years later, Schmitz returned to Sweden, hungry to make his own name as a scientist.

Earth's surface is pockmarked with impact craters left by large asteroids, so clearly the Chicxulub impact wasn't a rare event, even if that rock was especially gigantic and deadly. Schmitz reasoned that if one of these asteroids left behind it a tell-tale layer of material in the geological record, then it stands to reason that others might have too. His dream was to find 'the next KT-boundary'.

These thoughts were on his mind one day in the early 1990s, when he came across a short newspaper article. It concerned a man named Mario Tassinari, who lived on the slopes of a small mountain called Kinnekulle close to the town of Lidköping in southern Sweden. The article explained that Tassinari had found a fossil meteorite. To Schmitz, that was huge. He had hoped that there might be chemical and geological signs of ancient impacts from the largest asteroids – leftover traces of the vaporised rock – but to find an intact fossil meteorite was a jaw-dropper.

Schmitz would only learn it later, but Tassinari's discovery may not have been the first fossil meteorite find. In the 1950s, workers at a quarry in the Swedish county of Jämtland, in the centre of the country, found a strange black fragment of rock embedded in the limestone they were mining. It was thought to be nothing much until 1989, when two scientists re-examined it and showed that it was probably an ancient meteorite. Towards the end of the nineties, Frank Kyte at the University of California, Los Angeles, would delve into sediments taken from the bottom of the Pacific Ocean and discover a 2.5-millimetre-long fragment of rock. Given that the sliver of rock was found in sediments laid down at the time the dinosaurs died, Kyte wrote that he could infer this 'may be a piece of the projectile responsible for the Chicxulub crater'. Whether that is true is hard to say for certain. Kyte still has a piece of the fragment left, he told me, sealed securely in protective epoxy resin. Aside from these, there were only a few scattered and unreliable claims of fossil meteorites at the time Schmitz read that newspaper story.

Schmitz telephoned Tassinari straightaway and arranged to meet him. Tassinari lived in a slice of countryside nestled on the southern edge of Sweden's largest lake, Vänern. This area, called the Platåbergens Geopark, is home to no less than fifteen 'table mountains', so named because they have steep sides and flat plateaus on top. Tassinari had a house on the lower slopes of Kinnekulle, the smallest of the table mountains. With its pine forests, lake views and wild deer, it feels fresh and untamed.

It is also geologically fascinating, because the rock exposed on the steep hillsides is arranged in layers that go back millions upon millions of years in sequence. For those who know what to look for, it is a place where Earth's deep history is writ large, and you can wander along any trail and happen upon exquisite fossils of ancient sea creatures. Tassinari worked for a local museum and loved to collect these fossils. He also collected many other things – from old radios to corks from wine bottles – to the point that his home was so stuffed with specimens that it was hard to move around. Schmitz soon learned that, among his other geological treasures, Tassinari had indeed found what looked very much like a fossil meteorite. It had cropped up in a pile of waste stone at the Thorsberg quarry near Kinnekulle. This place would go on to be central to the study of fossil meteorites. In the summer of 2023, shortly after I had visited Jon Larsen in Oslo, I decided to pay it a visit.

I took a train across Sweden to Lidköping, where Schmitz picked me up in his car. It was early evening when I arrived, but Schmitz told me he wanted to take me straight to the quarry. It happened to be Walpurgis Eve, a public holiday on which Swedes traditionally build bonfires, and as we drove through the undulating countryside we saw no people, only neat pyramids of sticks waiting everywhere, ready to be lit. Perhaps it was just the fatigue from travelling, but as we neared the quarry and dusk drew in, it felt like we were a million miles from civilisation.

After slamming the car door shut, Schmitz walked with me to the quarry. We made our way over squelchy mud and past looming piles of waste stone, industrial buildings, water hoses and a huge circular rock saw twice as high as a person. The working rock face itself is smooth, pale and incredibly hard. It is the perfect stuff for the floor tiles of public buildings and indeed it is used as such throughout Sweden and elsewhere. That is why, years after he first met Tassinari, Schmitz would often find himself on his hands and knees in airports and railway stations, checking for a meteorite that had gone unnoticed.

As we stood in the quarry, Schmitz explained that when he first came here in 1994 he didn't think he would find more fossil meteorites – that still seemed unrealistic. Rather, he wanted to measure the iridium levels in the quarry to see if he could find spikes that would indicate the impact of another asteroid. 'I thought: this is where I am going to find the next KT boundary,' he says, gesturing towards the rock face. 'I thought I would be able to find several of these layers.' But when he made the measurements, he got a shock: the entire quarry wall had elevated iridium levels. It wasn't a spike – it was everywhere.

What that meant wasn't entirely clear at the time, but Schmitz struck a deal with the quarry workers. If they ever found a suspicious-looking smudge as they systematically cut up the rockface, they would give Tassinari a call, and in turn Tassinari would phone Schmitz. Once the quarry workers knew what to look for, they began to find three or four fossil meteorites every year. Schmitz analysed them at his lab in Lund University in the south of Sweden and confirmed what they were. Clearly, this little-known quarry was a unique place. It was almost the only source of fossil meteorites in the world and the best by a huge margin.

By the early 2000s, Schmitz and Tassinari had amassed dozens of fossil meteorites and a pattern was beginning to emerge. As we know, about 90 per cent of all meteorites found on Earth are of a type called ordinary chondrites. And of those, about 40 per cent are of a sub-type called L-chondrites, where the 'L' indicates that they are lower in iron than the others. It's fair to say L-chondrites are common. What Schmitz found, though, was that every single one of his fossil meteorites was an L-chondrite. That was puzzling: they were common, but not that common.

One possible explanation was that these meteorites were all chunks that had broken off from the same large meteor as it fell through the sky. But Schmitz didn't think that was plausible and, anyway, the ubiquity of the L-chondrites wasn't the only strange thing he found at the quarry. Schmitz also looked at the density of meteorites that had fallen over the area of the quarry, which allowed him to calculate a rough number for the flux of meteorites coming to Earth 466 million years ago, when the rock in which the meteorites were found had formed at the bottom of an ancient shallow sea. The flux of meteorites is not an easy thing to work out precisely, as we have repeatedly seen. But Schmitz's calculations suggested that the flux of meteorites 480 million years ago, during a period of history known to geologists as the Ordovician, was roughly 100 times greater than it is today. Even if that number was off by quite a bit, it still suggested that – for some reason – Earth had been getting a much more generous delivery of meteorites back then than it does now. But why? 'For me science is about curiosity,' Schmitz told me as we stood in the quarry in the gloaming. 'One question always leads to more questions – and I just want to know the answer.'

To get it, Schmitz needed more data. He was finding fossil meteorites in numbers that were unheard of, but there still weren't enough to produce statistically robust evidence of the increased flux that other scientists would accept. Still, he had a plan. The fossil meteorites he was finding were quite sizeable, perhaps the size of your palm. But as we learned when we met Jon Larsen, smaller meteorites fall more frequently and the really minuscule – micrometeorites – are more common still. Finding these inside the quarry rock was not going to be easy either, but Schmitz had a workaround.

The idea was to look for a mineral called spinel, which has a characteristic crystal structure in which the atoms are arranged in a repeating cubic lattice. Specifically, Schmitz wanted to find a type of spinel called chromite. As its name hints, this contains the element chromium, which is extremely rare on Earth but is guaranteed to be present in L-chondrite meteorites. Luckily, this mineral is much harder than limestone, the bedrock of the quarry. Schmitz took seven samples of rock, not just from the Thorsberg quarry but also from several different quarries in Sweden, amounting to more than 100 kilograms. He crushed each sample into a fine powder and dissolved away the limestone with hydrochloric acid. This left only grains of chromite, which could only have come from micrometeorites embedded in the rock. There were dozens and dozens of these grains, and by counting how many there were in rock formed at different times, Schmitz could reconstruct how the flux of meteorites had changed. The results, published in 2003, suggested a surge in numbers of incoming meteorites about 466 million years ago. The evidence was mounting.

Schmitz still wanted more. In September 2001, he attended the annual meeting of the Meteoritical Society, which was in Rome that year. There he met Rainer Wieler, an expert on the chemistry of extraterrestrial materials based at ETH Zurich, a research university in Switzerland. Wieler was known, among other things, for having the equipment and the technical know-how to work out the cosmic ray exposure age of meteorites.

The principle behind this technique hinges on the effects of cosmic rays, extremely energetic particles from deep space that continuously stream through the solar system. On Earth, we don't notice them because our atmosphere shields us, but in space they collide with lumps of rock and cause their atoms to change from one isotope to another. Imagine you have a huge asteroid orbiting out in space that collides with another rock and is smashed into fragments, some of which end up falling on planet Earth. Before that clash, most of the rock will have been buried in the interior of the asteroid and so sheltered from cosmic rays, but as soon as a chunk is liberated, cosmic rays start raining down on it and altering the balance of its isotopes. The stopwatch has started. Then, when those pieces reach Earth, they are under the protective umbrella of our atmosphere. The stopwatch stops. People like Wieler and his colleagues had learned to make the precision measurements that revealed this cosmic ray exposure age. They had the ability to measure how long any meteorite had been in space.

At the conference in Rome, Schmitz asked Wieler if he would apply his methods to the chromite grains. It was quite an ask. Calculating the exposure age typically involves heating a sample of rock with a laser and collecting the gasses that are given off. To try and do this on tiny grains of chromite would be fiendishly hard in itself. To expect that those gasses hadn't been disturbed by the fossilisation process seemed scarcely credible. Looking back on the time in a recent monograph, Wieler wrote that he was not convinced that it would be remotely possible, but Schmitz eventually won him over. 'Birger is a person of great enthusiasm and persuasion,' he wrote. 'So, we decided to give it a try, perhaps also just for fun.'

It was not only possible but it also revealed something hugely surprising. Wieler and his team measured the exposure ages of grains of chromite that had been extracted from rock of different ages in the Swedish quarries and they found that the exposure age of the chromite increased neatly and gradually depending on the age of the rock in which it was found. That might not sound so ground-breaking at first, but it is highly insightful. Some people had argued that Schmitz was finding so many meteorites because they were all really just bits of the same original meteor. Now that was off the table. If it were the case, the meteorites would all have the same exposure age. To the contrary, the fact that the meteorites' exposure age correlated with the date that they fell to Earth suggested that they were different meteorites that all originated from the same breakup event and had arrived here one after the other.

This was music to Schmitz's ears because it supported a hypothesis he had been nursing for some time. It went like this: roughly 500 million years ago, two huge rocks smashed together in the asteroid belt. One was what meteoriticists call the L-chondrite parent body; in other words, the asteroid from which at least some of those meteorites originated. We know this collision was a big one because the L-chondrites we find even today have evidence of the shock baked into their mineral structures. This huge collision created a cloud of meteoroids, some of which would eventually get sucked into an orbit that, millions of years later, crossed the orbit of Earth. Schmitz thought this pattern of events should have happened many times, so that the flux of meteorites reaching us would wax and wane over the course of history, as different asteroids broke up and sent waves of debris our way. What he was seeing in the Thorsberg quarry was, he believed, the result of one of these events: the breakup of the L-chondrite parent body.

Meanwhile, others had recognised that even if large fossil meteorites were almost impossible to find, extremely old micrometeorites might be more feasible. There had in fact been a few scattered reports of ancient micrometeorites going back decades. In 1991, Susan Taylor at the US Army Cold Regions Research and Engineering Laboratory and Donald Brownlee at the University of Washington in Seattle made a study of ancient meteorites locked in ice cores and ocean floor sediments, some of which dated to the Jurassic period. But this study and a few others like it didn't do much more than describe the stones. It is a technical point, but Schmitz also points out that these would not be considered fossil meteorites because they were encased in ice rather than rock.

Andy Tomkins heard whisperings of fossil micrometeorites and it ignited his competitive streak. 'I wanted to find the oldest ones,' he says. And why not? After all, Tomkins is based at Monash University in Melbourne, Australia – just a few hours' flight from the Pilbara region on the other side of the country, home of Earth's oldest known rocks. Looking there was surely worth a try.

In 2014, Tomkins and his student Lara Bowlt flew to Port Hedland on the coast of Western Australia, rented a four-wheel-drive vehicle and drove south for two hours into the desert. The scenery was all red rock and the traffic mostly road trains hauling containers across the interior. Eventually the road ran out and they drove along a dry riverbed, then hiked the last few miles to a site in Pilbara. 'It's beautiful, rugged country,' says Tomkins. They were headed for the Tumbiana formation, an area of rock formed at the bottom of a lake about 2.7 billion years ago. That takes us back to beyond the halfway point in Earth's lifespan. There was no guarantee they would find anything, but Tomkins and Bowlt dug out about ten kilograms of limestone blocks and had them shipped back to Melbourne.

There they cut the rock into small cubes and plopped them into a bath of hydrochloric acid for a day or two. That dissolved away the rock and, after sifting through the insoluble remnants with magnets and sieves, they ended up with sixty iron micrometeorites. That they were extraterrestrial was apparent from the presence of an iron oxide-based mineral called wüstite, which can only form at extreme temperatures of the sort produced when a meteorite encounters the scorching friction of Earth's atmosphere. Tomkins had achieved his goal. These were by far the oldest meteorites ever found.

The finds set Tomkins' mind whirring once more. Picture a shooting star streaking across the sky. It flashes bright and hot at first as it smashes into Earth's atmosphere at a speed of perhaps 72,000 kilometres per hour. Travelling at such high speed, the iron gets roasting hot and reacts with oxygen in the air to form wüstite. The rock then slows to a more leisurely pace and stops reacting in the lower atmosphere, before finally smacking into the ground. Tomkins realised that the iron in these meteorites had captured a sample of Earth's upper layer of atmosphere 2.7 billion years ago. 'It was the first time anyone had thought of a way to sample the upper atmosphere back in time,' says Tomkins.

The scientific consensus has long been that Earth's atmosphere contained more or less zero oxygen until 2.4 billion years ago. The uptick was possibly caused by the appearance of cyanobacteria, a type of microorganism that produces oxygen as a by-product of photosynthesis. But Tomkins' fossil meteorites seemed to give the lie to all that. When they zipped through Earth's atmosphere, Tomkins worked out that the concentration of oxygen ought to have been around 21 per cent – a similar value to that of today – in order to explain the creation of the iron oxide minerals. Tomkins and his team developed a model of the atmosphere that showed there could have been a separation that kept the upper and lower parts of Earth's atmosphere from mixing, meaning that any early oxygen wouldn't necessarily have been available to life on the ground. Still, it was a startling and bold claim.

When Rebecca Payne at Pennsylvania State University came across Tomkins' work, she was 'sceptical but intrigued'. She was convinced that these were genuine fossil meteorites and they had been oxidised – but she wondered if it was really oxygen that had been responsible. Working with her colleagues James Kasting and Donald Brownlee at the University of Washington in Seattle, she looked at whether it might have been carbon dioxide that had caused the oxidisation instead. The team had some experiments carried out in special experimental tubes, which briefly produce conditions that mimic the meteorites' roiling entry into our atmosphere. This showed that if the upper atmosphere contained about 25 per cent carbon dioxide, then that could have oxidised the molten iron in the way Tomkins thought oxygen had.

There was just one problem. Carbon dioxide is an extremely potent greenhouse gas. Today, carbon dioxide constitutes a fraction of 1 per cent of the atmosphere. If Earth was shrouded in an atmosphere made from a quarter carbon dioxide, it would have been a hothouse. That doesn't tally with what we know about the conditions 2.7 billion years ago – we know Earth had ice caps at the poles, for instance. Payne's conclusion was that the overall thickness of the atmosphere must then have been lower, at about 60 per cent of what it is today. 'We are fairly confident that carbon dioxide must be the oxidant for these meteorites,' says Payne. A team led by Owen Lehmer at NASA's Ames Research Center reached similar conclusions in a study published just after Payne's.

Tomkins now accepts that Payne is probably right and his initial interpretation was wrong. And he's fine with that. None of this diminishes what he's really excited about, which is using fossil meteorites as a way of probing the past. He has even done some work looking at whether Earth might have had a ring of dusty debris around it – a bit like the ones that encircle Saturn today – due to the deluge of cosmic dust it was receiving. At the time of writing, this work wasn't published, so details are scant, but it is an intriguing example of where the study of fossil meteorites can take us. Payne also thinks that these fossils can be instructive. 'These meteorites represent a part of the atmosphere that we would otherwise have no way to look at, it's lost to time,' she says. 'It's now a question of finding more.'

Martin Suttle, who has cropped up several times in this book already, was thinking along similar lines. In 2014, he had begun studying for a PhD in meteoritics at Imperial College London. Though he wasn't focused on fossil meteorites, he was so intrigued by the idea that he decided to try finding some in his spare time, so he went to a road cutting near his parents' home in Surrey in the south of England and collected some lumps of limestone. He bore them back to Imperial College and sent a shard of the rock for dating, which showed it was eighty-seven million years old. If there were meteorites to be found in this rock, they would have fallen during the late Cretaceous, the days of the vast sauropod dinosaurs, the largest animals ever to walk the Earth.

Suttle bought a large bottle of liquid pavement cleaner with which to dissolve some of the rock. This yielded ten micrometeorite particles. Then he tried grinding some other chunks to a powder and picking out the micrometeorites with a magnet. This yielded about another sixty. These tiny grains were very different from the chromite Schmitz had found. Meteorites are often marked out by the fact that they contain much higher concentrations of nickel than Earth rocks. These grains, Suttle found, were strange in that they contained the sorts of minerals you would expect in a meteorite, except the nickel had been replaced by manganese. At first that seemed inexplicable – except, actually, no it wasn't, Suttle realised.

Consider a fossilised dinosaur bone. Once buried in sediment, the animals' tissues begin to degrade and flowing groundwater often deposits minerals in their place in a process called diagenesis. The dinosaur 'bones' that palaeontologists find are not bone at all, but minerals deposited in the cavity where the bone used to be. Suttle reasoned that the same diagenesis process could have swapped the nickel in the meteorites for manganese.

This is a crucial finding because anyone looking through ancient sediments who came across particles like the ones Suttle found would not have thought of them as extraterrestrial because of the lack of nickel. 'We have probably not recognised fossil micrometeorites that might have been found by other people in the past because this fossilisation process alters their chemistry,' says Suttle.

He thinks they could be way more common than we previously realised.

Suttle went on to prove that he could find them in many other old rocks. He searched through samples of the ocean floor sediments brought back by the *Challenger* expedition, a scientific voyage of discovery in the 1870s that laid the foundations of modern oceanography. Micrometeorites had already been discovered in these samples, but Suttle managed to find previously overlooked ones. He also looked at rocks collected in the Ural Mountains that were formed during the Devonian period, roughly 400 million years ago. He found forty fossil meteorites in these samples too.

He then analysed a bunch of samples from Italy that were formed about eight million years ago, during the Miocene period. This is an interesting moment because scientists have previously inferred that Earth received a larger than normal amount of cosmic dust – that is, the micrometeorites that Jon Larsen collects – at that time. They can't find the dust motes themselves, but they do see a spike in helium-3, an isotope of the element which is known to be raised in cosmic dust. Suttle looked at rock samples from five time windows and dissolved twenty kilograms of rock from each, recovering about twenty micrometeorites in total. But there didn't appear to be anything interesting to see here: the flux of meteorites was flat. It isn't entirely clear what's going here, but one possibility is that there really was more incoming cosmic dust eight million years ago, but the extra material was not very hardy and was swiftly degraded away to nothing.

For David Nesvorný, it definitely makes sense for us to see occasional increases in the influx of cosmic dust, thanks to a mechanism called the Poynting–Robertson effect. This significantly affects dust but barely matters when it comes to larger meteoroids. Think of the effect as akin to radiation from the sun pushing on small dust particles in space in such a way that they spiral inwards towards the star – and so sweep across Earth's path on their way. This may explain why we sometimes see spikes in cosmic dust without an accompanying uptick in meteorites.

Back in Sweden, Schmitz wanted to take his search for chromite grains to an industrial scale. The idea was to build a new facility at Lund University dedicated to dissolving limestone with acid. That way, he would be able to properly investigate the chromite grains in rocks going back billions of years. He hoped it would be the birth of a new kind of stratigraphy, the branch of geology that studies rock layers going back in time. Instead of looking at biological molecules or animal fossils, as stratigraphers commonly do, he would be looking for materials from space. He called it 'astrostratigraphy'. In 2012 he got lucky: the European Union awarded him a grant worth €1.9 million to make it happen.

Schmitz worked on the project for years and he and his students dissolved more than 1,500 kilograms of limestone, amassing some 900 chromite grains. This was enough to reconstruct the flux of L-chondrites over tens of millions of years. Again, this showed an upwards snap in the number of incoming L-chondrites precisely 466 million years ago. Schmitz realised that this coincided with an event known as the Great Ordovician Biodiversification Event (GOBE). At this point in time, there was a major extinction, in which many species of animal died and new ones evolved to fill the gaps. We know that sea levels also dropped significantly at this time, which indicated that the temperature on Earth had fallen, locking away more water in the polar ice caps. Schmitz now thought he had an explanation for this. The incredible amounts of dust created in the breakup of the L-chondrite parent body would have swept over Earth and entered our atmosphere, reflecting away much more sunlight than usual and cooling the entire planet. In other words, the influx of L-chondrites and dust had cooled the planet and this had precipitated a great dying off of creatures. It was a hypothesis that stood alongside that of the Chicxulub impactor in terms of significance: this was a second occasion when rocks from space had profoundly influenced the course of life's evolution.

Since Schmitz started his research on fossil meteorites, scientists have become much more divided on the exact timing and extent of the GOBE. As a result, he now tends to avoid talking too much about that ill-defined extinction event and instead relates the spike in fossil meteorites to the mid-Ordovician ice age, a much more robust and well-accepted idea. Apart from sorting out these details, Schmitz also wanted to investigate a larger question to do with the way meteorites are delivered to Earth. The basic hypothesis for how this works is known as the 'collisional cascading model'. It says that, from time to time, one asteroid would hit another and the collision would release a barrage of smaller rocks. These would then cascade into the inner solar system, some hitting Earth. Since these asteroid collisions would happen from time to time, there ought to be successive waves of meteorites hitting Earth as a result, so you would expect occasional peaks in the flux. This is exactly what Schmitz had observed for the L-chondrites – the question was, were there other such peaks corresponding to other asteroids breaking up?

Schmitz reasoned that it should be possible to tell. Just as L-chondrites contain those chromite grains, some kinds of meteorites from other asteroids contain different kinds of hardy spinel minerals. Looking at other spinels would open up a window on at least fifteen different types of meteorite, originating from two different kinds of parent asteroids (the S-types and the V-types). To hunt for them, Schmitz and his colleague Fredrik Terfelt took stone samples from quarries that represented fifteen time windows spaced throughout the past 500 million years. It amounted to more than 8,400 kilograms of rock in total – and they dissolved the lot of it, to find grains of spinel mineral. But apart from the spike in spinels from the L-chondrites, there were no other spikes at all related to any other meteorite. For Schmitz, it was a deeply unexpected result. 'How can you explain just one type of meteorite coming in for 500 million years?' he asks. This result has shifted his perspective on how meteorites get to Earth. He still thinks that asteroids must break up fairly often, but that it must be rare for the debris created to be sent Earthwards. So there must be some highly unusual reason why the L-chondrites bucked the trend. 'I think there has to be some new mechanism we haven't thought of yet,' he says.

Bill Bottke at the Southwest Research Institute in Boulder, Colorado, is deeply impressed with all this. 'Birger's work is spectacular,' he says. Bottke is an expert on asteroids and is known for his work on trying to deduce the families of asteroids from which the various types of meteorite originate. It is a knotty problem, not least because there are so many asteroids out there. You can build up several lines of evidence that all point to a small group as the probable source of some type of meteorite, but it's very hard to prove the case, because that would mean eliminating the scores of other possible candidates.

There are only a few cases where we have pretty solid evidence tying a kind of meteorite to a family of parent asteroids. One of these happens to be the L-chondrites. According to Bottke, several lines of evidence suggest that they came from the stony Gefion family of asteroids and this would be his best bet as to the L-chondrites' parent body. Assuming that is true, the question is why, when Gefion broke up, did that produce a rain of meteors on Earth when other asteroids don't seem to have done the same. Why is Gefion special?

Bottke has an answer. To get to grips with it, we need to first understand how a space rock gets moved from the asteroid belt to Earth. This process is thought to be governed by something called orbital resonances. Imagine that the solar system is like a modified clock face, with the sun at its centre. This clock has one hand that traces the movement of an asteroid: this is short and moves fairly quickly. It also has another hand that traces the movement of Jupiter: this is longer and moves more slowly. Here's the vital part: these two hands move at different speeds, but at regular intervals they line up. It turns out that the real situation in the solar system is just like this: things orbiting at certain selected distances from the sun do so in a way that leads to repeating alignments – a state that astronomers call being in resonance. This means that Jupiter's incredible gravity gets the chance to repeatedly tug on some – but not all – of the asteroids in the belt. This can upset their orbit with wild consequences, sometimes in such a way that they are sent careening into the inner solar system and towards Earth. This isn't just speculation. Astronomers have spotted a series of empty rings in the asteroid belt, which are called the Kirkwood gaps. These rings match up in terms of resonance and so it seems they have been cleared of asteroids because any that stray into these no-go zones are pulled out by Jupiter's gravitational heft. All this makes a huge difference to what meteorites come our way. Any asteroids or meteoroids that are nowhere near a Kirkwood gap are likely to remain where they are and have little chance of ever coming near Earth.

Let's add some more detail to the picture. When an asteroid gets smashed up, Bottke says, it immediately creates a cloud of debris with lots of big rocks. But over time, the rocks pepper each other with impacts and are eventually ground down to a powder so fine it has no hope of reaching Earth. This means you have a limited time window during which meteoroids need to be tugged by one of those orbital resonances. It turns out Gefion is right on the edge of a 5:2 resonance (so, for every five orbits of the asteroid and every two orbits of Jupiter, the two bodies line up). In other words, Gefion broke up in exactly the right place. 'It just happened to break up near a superhighway,' says Bottke.

Other asteroids that break up might simply be too far away from a resonance zone to ever get pushed towards Earth. Fragments from such events might eventually get there, but Bottke reckons such events might end up delivering a lot less material to Earth, to the point where it is difficult to recognise in the terrestrial geological record. He does think there could be other asteroid collisions which happened close to a resonance – and maybe one day we will find evidence of one. 'It's a question of finding exactly the right outcrop of rock – and that's difficult.'

That may mean that the Thorsberg quarry will remain a one-of-a-kind place. For Schmitz, of course, it will always be special. These days, he doesn't spend quite as much time on his hands and knees as he used to, but he always keeps one eye open for fossil meteorites on likely-looking floors. As we drove back from the quarry through the Swedish countryside, I asked him if he had ever found one outside the quarry.

He had actually found two. One was in the floor of a small train station in central Sweden, tucked out of sight behind a staircase. Schmitz reckons the builders probably thought the stone had a defect. 'To them,' he says, 'this was a second-grade rock.' The other was on the steps of Gunnebo House, a grand residence on the outskirts of Gothenburg that the public can visit. He noticed it one day and spent some time considering it, but the next time he visited, a few years later, it had disappeared, the wind and rain having weathered it away to nothing.

CHAPTER NINE

Where Meteorites Come From

THERE IS A brown envelope waiting for me on the doormat when I get home from work. I can see from the postmark that this is the package I have been expecting. I slide my finger under the seal and extract a plastic zip-lock bag. The question is how badly are the contents going to stink? I open it just a tiny bit and, gingerly, take a sniff.

It is not as bad as I had feared. Still, I find it difficult to inhale a whole lungful of air without it sticking in my throat. There is a smokiness to it, a subtle acridity. It is the sort of smell you might catch if someone burned a tyre in the woods and you walked past the next day after it had rained. But if I'm honest, it is hard to do this smell justice with words. And no wonder – there is nothing else like it on Earth. I have been sent a postcard impregnated with the smell of a comet.*

* The human nose is a fascinating and varied thing. Not everyone has the same set of olfactory receptors, which means you might not experience the smell of the comet in the same way I did. To test this out, I asked my wife to sniff the postcards without telling her what the smell was. 'Hmm, I quite like it,' she said. 'Herbal. Earthy. Relaxing. It smells like something I would put in a hot bath.' Make of that what you will.

In 2014, scientists behind the European Space Agency's Rosetta mission landed a probe on a ball of dust and ice called comet 67P/Churyumov–Gerasimenko. It was all over the news at the time and rightly so. We have been sending spacecraft to other heavenly bodies, like the moon, Mars and beyond, for years, but comet 67P is tiny in astronomical terms, at just over four kilometres across. To land a machine on it and measure its properties was quite the feat. As we know, the probe measured the comet's deuterium–hydrogen ratio, but we also got to know its overall chemistry so well that scientists could reconstruct what it would smell like, devise a cocktail of molecules to represent that odour and then impregnate it into a set of postcards.

Comet 67P isn't the only small astral object we have called on lately. Over the past twenty years or so, we have sent probes to at least five small asteroids and comets drifting out in the solar system and brought morsels home. The thrill of discovery is ample reason to do this – who knew that comet 67P would look like a cosmic kidney bean, for instance?

But there are also vital scientific reasons for wanting to visit the parent rocks that give birth to meteorites and bring samples back to Earth for study. For one thing, there is always the chance that a dangerously large asteroid might be on a collision course with Earth in the future. We don't believe that any asteroids that are at risk of crossing Earth's orbit are big enough to cause a global extinction event, but there are some so-called 'city killers' that would cause a regional disaster if they hit the Earth. One is the 500-metre-wide boulder Bennu, which we will meet again later, and which has the highest probability of colliding with us out of all known asteroids. Scientists have calculated it will pass between the Earth and the moon in 2135 and that's when there is a small chance that things could get messy. (It's not as far off as you might think. I'll be long gone, but my grandchildren could live to find out what happens.)

If we did get wind of one of these asteroids bearing down on us in the centuries to come, we would need to work out how to stop it. Scientists are already weighing up the options, from simply blowing it up to the exotic-sounding 'gravity tractor' technique, which involves flying a craft alongside the asteroid and using its small gravitational pull to slowly tug the incoming rock onto a different trajectory. The best course of action, however, depends on the asteroid's composition and consistency. Blowing it up won't work if it just creates lots of slightly smaller rocks that would still hit us and do a lot of damage. So, it helps to go and see them and find out what they are like.

Many asteroids are replete with rare, valuable metals that some reckon we might one day harvest in space mining operations. In principle, there is an advantage to this because mining on Earth is frighteningly destructive to our environment, yet we desperately need an array of metals in things like batteries, one of the technologies that underpin our race to wean ourselves off fossil fuels. If we could collect such metals from asteroids instead of mining them on Earth, on the face of it that could be a good thing. Amazon founder Jeff Bezos is a fan. In 2021, he told NBC news: 'We need to take all heavy industry, all polluting industry, and move it into space and keep Earth as this beautiful gem of a planet that it is'. This is why many asteroid mining proponents are interested in getting the ground truth about what asteroids are really like – to lay the groundwork for resource extraction business ventures. Personally, this leaves a nasty taste in my mouth. I think it will be so costly and challenging that the business case will never add up. Better to concentrate on reducing our appetite for metals and slashing the impact of mining down here on Earth.

The most important reason to visit asteroids is for the purpose of fundamental science. These space boulders are where meteorites originally come from, and as such they are the ultimate target for the extraterrestrial rock hunter bent on understanding the history of the solar system. Burgling samples from asteroids can fill in the gaps in the knowledge we have gleaned from meteorites. For one thing, as we already know, the space rocks that fall to Earth are not a representative sample of what's out there in the solar system. We get only the meteorites that pass near us. Plus, even the ones we do receive get chemically changed in ways we don't fully understand as they plummet through our atmosphere.

Most of all, these missions can help us understand which types of meteorites come from which kinds of asteroids. It's not that we have zero knowledge about this. We can reasonably infer that stony meteorites must come from stony asteroids, whereas the damp, dark carbonaceous chondrites probably come from damp, dark parent rocks. But it is challenging to go further and match specific groups on the meteorite family tree with their parent asteroids. True, we can look at the light reflected by asteroids to get a sense of what they are made of, and this can assist in match-making.

But a confident judgement depends upon comparing the chemistry of the meteorite with the asteroid – and for that, we must go and get some of the latter.

It's not all fun and games. As the saying goes: space is hard. The prize at the end, however, is worth the effort. By tracing meteorites to their parent asteroids, we are creating a kind of time-ordered geological map of the solar system that can show us how rocks moved around. This could deepen our burgeoning understanding – on display in the Nice model, for example – of how the solar system grew to be as it is today.

The first mission to do a 'sample-return' – to take a piece of a heavenly body and bring it home – was Apollo 11, which famously took Neil Armstrong and Buzz Aldrin to the surface of the moon in 1969. That was closely followed by Luna 16, a Soviet mission in 1970 that used a robotic lander to return about 100 grams of regolith, or the moon's 'soil'. All in all, before the Apollo programme was cancelled in 1972, its missions brought back approaching 400 kilograms of lunar rock and dust.

For Mike Zolensky, curator of astro-materials for NASA, one of the most important early sample-and-return missions was the Long Duration Exposure Facility, better known as LDEF – and it never even went to a heavenly body. This was one of the first missions to be flown on the US Space Shuttles when they started in 1981. LDEF was a twelve-sided metal cylinder the size of a bus, with trays fixed to its sides. The idea was to test how various objects from Earth fared when put into space for long periods. For example, some trays contained bacterial spores (about a third of which survived their ordeal) and others contained tomato seeds.* But many of the trays were designed to catch the tiny micrometeorites zipping past in space. This had never been done before. Scientists wanted to know if it was possible and, if so, what these tiny meteorites were made of.

LDEF was placed in orbit by the *Challenger* Space Shuttle in 1984 and the original plan was to nip back and collect it within a year, but a series of delays – in part because of the fatal explosion of the *Challenger* – meant it was left in space until 1990, when another space shuttle collected it. When Zolensky and his colleagues finally got their experiments back, they found there were indeed lots of tiny micrometeorites embedded in the trays. They had pulled off the feat of catching space dust before it hit the Earth. As well as being an impressive demo, this helped quell the lingering uncertainty about what kind of material there was swilling around our planet. The satellite had its trays pointing in different directions, meaning some mostly caught particles zipping straight down towards Earth while others caught more of those in orbit. By looking at tiny impact marks, Zolensky and his colleagues worked out how much dust there was in orbit and how fast the micrometeorites were travelling. A few years later, in 1996, scientists launched a

* A total of 12.2 million tomato seeds were sent up on the LDEF mission. Once they returned to Earth, they were distributed to millions of schoolchildren across the US and grown in a mass experiment to check if being exposed to the radiation in space affected the seeds' growth. According to a NASA press release entitled 'Attack of the killer space tomatoes? Not!' there were no physical differences between the space tomatoes and ordinary ones – though the space tomatoes did tend to grow slightly faster at first.

similar micrometeorite collection experiment called the Orbital Debris Collector, this time hanging it on the Russian Mir space station. But by this time, scientists were itching to try something far more ambitious: getting a sample of the sun.

To see why, it helps to know that around the turn of the twenty-first century the science of meteorites was even more baffling than it is now, and part of the reason has to do with our old friends the isotopes. This time, the issue was around isotopes of oxygen. All the stuff on Earth – the air, the rocks, the trees – has a characteristic mix of oxygen isotopes. There are three main players here: oxygen-16, oxygen-17 and oxygen-18. I'll spare you the fine details and just say that there is the same predictable relationship between the relative proportions of these three oxygen isotopes in all natural materials on Earth. Scientists can plot the measurements on a graph, giving us what's called the 'terrestrial fractionalisation line'. Measure the oxygen isotopes on any Earthly rock and they would fall on this line.

Yet, in the 1970s, meteoriticists had realised that meteorites did not dance to the same tune. If you measure the oxygen isotopes in meteorites, they don't fit neatly on the fractionalisation line, but instead they sit in isolated clusters to either side. We interpreted this to mean that each group of meteorites comes from a different parent body, with each cluster on the chart representing the characteristic mix of oxygen isotopes found in a different asteroid family.

You might legitimately wonder how can this be? Those parent rocks would have first formed from the solar nebula, that disc of dust and gas that encircled the young sun. This material was swirling around under the influence of gravity, so you might think that, just like when you beat a cake batter, all the ingredients should have been evenly mixed, with the concentrations of isotopes the same everywhere.

But there are good reasons to doubt that this remained true for long. Take carbon monoxide, a simple molecule made of one carbon atom bonded to one oxygen atom, which would have been present in that dusty primeval disc. As sunlight shone on these molecules, they would have absorbed the energy and split into the two constituent atoms. The precise wavelength of light needed to break apart the molecules depends on the isotopes from which they are made. When it was first formed, the disc is thought to have contained more carbon monoxide-bearing oxygen-16 and this would have filtered out particular wavelengths of light, preventing it from reaching the edges of the solar system. The upshot? Well, the relative amounts of the oxygen isotopes in different parts of the disc changed, and so asteroids forming at different distances from the sun are isotopically distinct. Those same differences remain present in meteorites today. It isn't the easiest thing to get your head around, granted, but all this is exciting for the meteorite expert, because it means measuring isotopes can potentially tell you where in the solar system a meteorite's parent body originally formed.

The only trouble is, for this to work well, you need some solid reference point to work from: you need to know what the ratio of those oxygen isotopes was at the year dot, when the sun had only just formed. Dauntingly, the only way to measure that would be to sample the sun itself. It was the first thing to form in the solar system and it is assumed to have a composition that matches the original disc of dust.

I'll level with you, scooping a sample of the sun is not going to be possible any time soon, but luckily our star constantly sheds bits of itself and sends them flying out into space. This constant deluge of charged particles and ions is called the solar wind – and Donald Burnett had a plan to catch some of it. Burnett, who is based at the California Institute of Technology in Pasadena, came up with the idea for a NASA mission called Genesis. When he was developing the idea, Burnett told me, the best estimate people had for the solar isotope composition came from meteorites themselves, because those were the oldest materials we had. This had irked him for a long time. It surely wasn't a good idea to use a measurement from a meteorite as a baseline from which to make judgements about other meteorites – for a scientist, that was uncomfortably self-referential. 'I had basically been wanting to do Genesis my whole life,' Burnett says.

The mission launched in 2001 and travelled to Lagrange point 1, a spot between the Earth and the moon where two large gravitational forces cancel each other out and a craft can easily remain for long periods. Here, *Genesis* was beyond the influence of Earth's magnetic field, which would otherwise deflect away the solar wind. Then the craft opened up like a pocket watch and exposed two circular panels fitted with 250 kinds of wafer designed to collect particles from the solar wind. In a similar way to LDEF before it, these wafers were designed to absorb the particles, slow them down and hold onto them. The wind is mostly protons, and bears only trace amounts of isotopes of elements like oxygen, nitrogen and carbon. To collect enough, *Genesis* had to hold position for about three years. It only released its sample capsule in 2004, which then began its descent back to Earth.

That was when things went awry. The craft's delicate wafers were extremely fragile, so Burnett's plan had been for the capsule to drift down on a parachute and for a helicopter to swoop in and deftly snare it. The operation had been practised over and over. However, it turned out that a decelerometer on the capsule had been installed backwards by mistake and the parachutes were never triggered. The helicopter pilots didn't stand a chance, as the capsule shot downwards at more than 300 kilometres per hour and slammed into the desert floor in Utah. 'One of the problems was that the return was not very graceful,' says Burnett. 'When you look at pictures, it looks as if the capsule is buried in the ground, but actually the ground is much harder than the spacecraft and it is just crunched up – it was a big mess.'

However, he and his team did manage to retrieve some relatively undamaged pieces of wafer from the wreckage and analyse the solar wind particles (it took a little longer than anticipated, as they had 15,000 tiny pieces to work with instead of 250 wafers). Among other things, the measurements showed that the solar composition had far less oxygen-18 than expected. Ever since, the baseline numbers from Genesis have been used as a crucial part of the way scientists model the evolution of the solar system's geology.

Around the same time as *Genesis* was silently bathing in the solar wind, scientists and engineers in Japan were thinking about how they could make their mark on solar system exploration. The Japanese Aerospace Exploration Agency (JAXA) did not have the resources of NASA or the European Space Agency (ESA), so a mission to the moon or Mars, say, seemed too ambitious at the time. But visiting an asteroid and returning with a souvenir seemed doable – and it would be an impressive first. 'We felt that sample-returns were the only path on which we could be visible,' says Masaki Fujimoto, deputy director general of JAXA's Institute of Space and Astronautical Science.

In 2003, JAXA's spacecraft *Hayabusa* (the name translates as 'Peregrine Falcon') took off and began its journey to an asteroid called Itokawa, which had been chosen, Fujimoto says, 'because it was easy'. Itokawa, which is just over 500 metres long, follows an orbit that takes it swinging in towards the sun and then outwards on a path that comes close to Earth. It is an S-type asteroid, which is common: 17 per cent of known asteroids are of this variety. Based on what we can tell about their make-up from the way they reflect light, it was thought that these asteroids have a similar composition to stony meteorites. It would make sense if the various types of stony meteorites came from different S-type asteroids.

Itokawa may be accessible as asteroids go, but getting around in space is never a piece of cake. Because space offers no resistance, craft typically need to shoot out propellant to both speed up and slow down, but they have a limited supply of that propellant. This is why space missions often rely on the gravity of large objects like planets to help them manoeuvre without using up fuel. With Itokawa, though, the JAXA scientists were trying to approach a tiny object with very little gravitational pull. Because of this, they had to send *Hayabusa* on a long path that approached Itokawa gradually. They also used a specially designed type of ion engine, which could work for longer than ordinary rocket propellant.

Hayabusa turned out to be a rollercoaster of a mission where more or less anything that could go wrong did go wrong. First, as the craft was on its way to Itokawa, it was hit by one of the biggest solar flares in recorded history, a massive burst of radiation spat from the sun. This damaged the craft's solar panels, which in turn reduced the power of the ion engines. In autumn 2005, behind schedule, it approached to within twenty kilometres of Itokawa and began filming it, revealing it was shaped like a kidney bean. *Hayabusa* closed in and released a small lander called MINERVA that had been designed in collaboration with NASA to hop along the asteroid's surface and take pictures. Unfortunately, it missed the asteroid entirely and floated away.

Hayabusa then attempted to land on the asteroid for the first time. But disaster struck again: it stayed close for too long and the surface of the asteroid, heated by the sun, warmed the craft too much – kicking in a mechanism that saw the probe automatically retreat to a distance of 100 kilometres to preserve its delicate systems. JAXA engineers slowly regained control of the craft, which had been put into a wild spin by its retreat, and then tried a second approach.

This time, things seemed to go well. The craft descended and fired a small metal ball at the surface designed to stir up dust that would flow into its sample chamber. I do enjoy the thought of this plucky, if accident-prone, robot blowing a pea-shooter at an asteroid. Surprise, surprise: something went wrong again and JAXA lost contact with the craft. As the agency's engineers struggled to re-establish communication – for hours, days, then weeks – it was unclear whether any sample really had been grabbed and whether the probe was still functioning at all.

Meanwhile, back on Earth, scientists were gawping at the incredible images and measurements the craft had sent back. In June 2006, the prestigious journal *Science* devoted an issue to Hayabusa and chewed over what it had found. The completely unexpected thing was that Itokawa tuned out to be not so much a rock as a pile of rubble. No one had expected this, because with such weak gravity it wasn't clear how something Itokawa's size would persist in space unless it was stuck together quite strongly. 'It is not even clear why Itokawa is there at all, given that you just have to shake it gently [...] for it to fly apart,' wrote planetary scientist Erik Asphaug at the University of Arizona at the time. It is now thought that many smaller asteroids are rubble piles. That represents an about-face in the way we think about these bodies – and one that will be important if we ever have to deal with a massive one on a collision course with Earth.

Eventually, JAXA mission specialists re-established communication with *Hayabusa* and it limped home. In 2010, it entered Earth's atmosphere and its sample capsules slammed into the Australian desert. In the end, the capsules did contain about 1,500 tiny grains of material from Itokawa. Analysis confirmed that this material was more or less identical to LL-chondrites, one of the most common types of stony meteorite. It was rare solid evidence linking a particular type of meteorite to a particular type of asteroid.

Inside JAXA, discussions about a follow-up to Hayabusa started even while the first craft was in space and encountering all those problems. As it became clear that the first mission had been a success – albeit a tortured one – the atmosphere became more supportive in Japan, says Fujimoto, and a second mission got the go-ahead from the Japanese government. This time, the plan was to visit a more interesting asteroid, one that came from the dawn of the solar system.

Hayabusa 2 shot into space in 2014, destined for the asteroid Ryugu.* From looking at Ryugu's spectra – the particular wavelengths of light it absorbs and reflects – scientists suspected it was a C-type asteroid. These dark-coloured asteroids are thought to be the parent bodies of the rare carbonaceous chondrites (such as the Winchcombe meteorite) that contain water and organic molecules, the ingredients for life. As the potential origin of these special treasures, Ryugu was an attractive destination.

Although this new target is chemically very different to Itokawa, and significantly larger, at about 900 metres

* Ryugu is named after the underwater palace of a dragon god in a Japanese fairy tale. In the story, our hero is taken to the palace by a turtle he saves on the beach and is there entertained for several days by a princess. He is then sent home with a mysterious box that he is forbidden from opening – but inevitably does. *Hayabusa 2* would eventually return from Ryugu with its own box filled with mysterious contents. (In case you are wondering, Itokawa is named after a famous Japanese rocket engineer.)

across, it has a similar orbit that roughly follows that of Earth. The journey went smoothly this time and *Hayabusa 2* arrived at Ryugu in the summer of 2018, whereupon it was discovered that the asteroid is shaped like a squashed sphere. This second JAXA asteroid mission was more sophisticated and the orbiting craft dropped no fewer than three rovers onto the surface, all of which hopped and tumbled around making various measurements, including the porosity of the asteroid and its magnetic field. In 2019, the *Hayabusa 2* craft itself also swooped down and collected some material from the surface.

Then came the real showstopper. The JAXA scientists had also come up with an audacious plan to excavate material from the interior of the asteroid, in order to compare this with the surface rock they already had. To do this, they had essentially equipped *Hayabusa 2* with a bomb. A sort of space cannon was detached from the main craft, which then took cover on the other side of the asteroid. The cannon then fired a copper ball crammed with 4.5 kilograms of plastic explosive at Ryugu – which promptly exploded, creating a ten-metre-wide crater. The main spacecraft then emerged from hiding to scoop up some of the newly exposed subsurface material. Job done, *Hayabusa 2* fired its ion engines and headed homewards, eventually dropping off its sample containers in 2020.

The five grams of pristine Ryugu inside the containers held two insights that wowed scientists. To get to grips with the first, it helps to know that, although Ryugu, as a C-type asteroid, was always assumed to be more interesting than Itokawa, these kinds of asteroids are not actually rare. They are the most common type, with about 75 per cent of known asteroids thought to be C-types. Ryugu wasn't deemed to be a special one; it was simply picked because it was accessible. But when the particles from Ryugu were analysed, it turned out to be a close match for the CI-chondrites, which are one of the rarer sub-types of carbonaceous chondrite. In other words, this common or garden asteroid turned out to probably be the source of just about the rarest meteorites going. 'We expected Ryugu to just be a typical asteroid,' says Fujimoto. 'So, what does this mean, did we just get super lucky?' For his part, he says that the trip to Ryugu has underlined again how the meteorites we get on Earth are not at all representative of what's really out there in space.

The second insight connects to one of the most profound questions about the history of the solar system, and specifically its largest planet, Jupiter. To see how scientists arrived at it, we need to delve into a concept known as the snowline. Picture again the solar nebula, the disc of gas and dust of the early solar system. The solid particles slowly clump together to form rocks and boulders, but what is solid and what is gas? Well, it differs depending on *where* you are in that disc: as you move further from the sun it grows colder, so materials with successively lower boiling points are solid. This means that when you get about as far away from the sun as Mars is today, you cross a line after which solid water ice would be present. Go further and you begin to get solid carbon dioxide, further still and ammonia solidifies – and so on. These are the solar system's snowlines. By looking at what's present in Ryugu and what's not, scientists could work out where it formed. The fact that they found carbon dioxide buried inside its crystals showed that Ryugu must originally have formed in the outer solar system, roughly around the present orbit of Saturn. Isotope measurements, now reified thanks to the Genesis mission, also confirmed Ryugu's birthplace.

That in itself was not so unexpected. But the researchers were also able to deduce from the mineralogy of the Ryugu samples that it must have moved inwards, closer to the sun, within about five million years of the formation of the protoplanetary disc. This points strongly to there being some powerful gravitational force that pushed Ryugu inwards. What could this have been? According to Fujimoto, one good explanation would be Jupiter. If true, that indicates the gas giant was well on its way to its current size very early in the history of the solar system – another piece of evidence supporting the Grand Tack model we encountered in Chapter Six. 'This connection to Jupiter is a beautiful result,' says Fujimoto.

JAXA isn't the only agency with a dog in this race. NASA also has an asteroid sample-and-return mission called OSIRIS-REx, which visited the asteroid Bennu – the one that might hit us in 2135 – and returned its sample in September 2023. Bennu is another C-type asteroid, so it will be fascinating to see what kind of meteorites it matches once scientists have finished analysing the material. Will it be the more common kinds of carbonaceous chondrites or rarer ones, like the CI-chondrites that Ryugu matched with? Finding out will give us a better sense of whether we were just lucky that Ryugu maps on to a very rare type of meteorite, or whether there is something else going on. And they certainly have a lot to work with: OSIRIS-REx brought back about 120 grams of material, the largest amount of asteroid we have by a country mile.

After dropping off its sample containers on Earth, the *Hayabusa 2* spacecraft has been sent on an extended mission to rendezvous with another two asteroids, one in 2026 and another in 2031. It won't take samples of these – only images – but the Japanese taxpayers are certainly getting their money's worth from that mission. At the same time, the agency is planning another mission called MMX (for Martian Moons Exploration), which is scheduled for launch in 2024. Part of the plan is to land a craft on the larger of Mars's two moons, Phobos, take a sample and bring it home. That might sound like a departure from asteroids. But based on what we can judge about Phobos' geology, some scientists think that it started life as an asteroid that was captured by Mars's gravity. If true, this means that, like Ryugu, Phobos is another small body that pinballed around the early solar system. 'We are assuming that Phobos is a captured asteroid that formed in the outer solar system – but we don't know,' says Fujimoto. The MMX mission should tell us.

You could reasonably argue that this is a special time when it comes to asteroid exploration. 'We are living in the golden age of sample-and-return, for sure,' says meteoriticist Jemma Davidson at Arizona State University. Her colleague Lindy Elkins-Tanton is leading yet another mission in partnership with NASA, which launched in October 2023, to an asteroid called Psyche. This one is particularly compelling because the asteroid is probably a lump of pure metal. Based on the way it reflects light, we believe it to be the exposed iron–nickel core of what was once a small planet. There were presumably once many of these 'planetesimals' that didn't make it into proper planets, either because they ran out of material to draw in or because they were smashed to pieces by a collision. Some, however, would have got big enough for their own gravity to suck the iron in their constituent rocks into a central core, a kind of precursor to the liquid metal core that Earth has today. It is thought that Psyche is the remnant of one such metal core. The mission won't actually retrieve a sample, but we have never studied anything like this before, so taking a look will be a fascinating first.

For many planetary scientists, the next logical step on this path would be to design a sample-return mission that steals a piece of a comet. Meteorites are wonderful time capsules that tell us a lot about the early solar system, but the material in them has, to a greater or lesser degree, been heated and chemically modified over the course of their long lives. Comets are thought to contain material that has always remained frozen and unchanged, making it a pristine molecular record of the solar system's starting state. This is because comets follow off-centre and highly elliptical orbits that keep them far away from the sun's warmth nearly all the time. (Imagine looping a rubber band over your finger and pulling it taut. Comet orbits are like that: only passing the sun – represented by your finger – briefly at one end of their orbit.) Among other things, it would be fascinating to know what kinds of carbon-based molecules comets contain because that would tell us what molecules have been available throughout the solar system to be built into living things. 'Comets give you an unprocessed history of the protoplanetary nebula,' says Alexander Hayes at Cornell University.

We have already sent missions to comets, notably the Rosetta mission, which resulted in those funny-smelling postcards that dropped onto my doormat. Even before that, in 2004, NASA's Stardust mission visited a comet. But both focused mostly on the comets' coma, or tail, comprised of material subliming into space. We have never sampled the icy heart of a comet – but Hayes is out to change that. He is leading the development of a mission concept called CAESAR (for Comet Astrobiology Exploration Sample Return). The idea is to visit comet 67P, or possibly another comet, depending on the launch window, nab a sample of its core and let this warm up and evaporate into a gas canister. As long as that canister is sufficiently large, the pressure will remain low and the molecules from the comet will remain chemically unaltered.

One of the big questions in astrobiology, Hayes says, is what kinds of organic molecules would have been available on planets soon after they were formed, because they were already in the solar nebula – and what would have to form on the surface of a planet. Until you have this detailed understanding of the pre-biotic chemical possibilities in the solar system, you can't rule in or out where life could have formed, whether that be Mars or some exotic moon in the outer solar system. CAESAR could give us the information we need.

Hayes and his colleagues put CAESAR forward to compete for NASA funding in 2017 as part of the agency's New

Frontiers programme. It was selected as one of two finalists, but the other contender won. (To be fair, the other mission, called Dragonfly, involved sending a robotic helicopter to Titan, the wild moon of Saturn where there is rain and there are even lakes made of methane – who could compete against that?) Hayes is undaunted and is still developing the technology for his mission, in order to be fully ready for the next opportunity.

Space scientists also dream about one day sending a mission to a comet that would bring back the ultimate prize: a frozen, solid sample. That would mean preventing the comet ices from melting, and so require a container that could keep them extremely cold, at perhaps −150 degrees Celsius. This device would have to run for years, possibly on meagre solar power, and even keep its cargo safe during a blistering re-entry into Earth's atmosphere. No one is entirely sure how to make this bullet-proof space freezer at the moment. But I think we will succeed sooner or later. And perhaps it is then that the hunt for the ultimate extraterrestrial treasure will begin.

EPILOGUE

Back in the Basement

I WROTE MOST of the first draft of this book in autumn 2023. That turned out to be bad timing in one respect, because quite a few of the key researchers I wanted to speak to were preoccupied with a historic event: the return of the OSIRIS-REx mission, which was bringing back a sample of the asteroid Bennu.

At the time, I barely had a chance to follow the drama; I was shut up in my office typing feverishly. But when I returned to my normal magazine work a few months later, things had quietened down for the OSIRIS-REx scientists, and I managed to bag an invitation to speak to Sara Russell, one of the key researchers analysing the Bennu samples. I went over to see her at the Natural History Museum and found myself back where this whole saga started: in that basement room where Martin Suttle had introduced me to meteorites.

Now the floor-to-ceiling ranks of drawers with their meteorites inside had a new companion. In the corner stood a shiny glove box, one of those air-locked glass cabinets for handling sensitive materials. Russell told me it had been purchased after the fall of the Winchcombe meteorite and was now also being used to protect pieces of Bennu from the degrading influence of Earth's atmosphere. It was a symbol of how much had happened in the world of meteoritics in just a few short years.

Russell laboriously inserted her fingers, hands and arms into the rubber gloves, reached inside the glove box and then unwrapped a plastic bag containing a small chipping stolen from Bennu. She held it up for me to see. I pulled a cliched journalist move and asked how it felt to hold something so old and precious. 'Every time I handle it, I'm absolutely terrified,' she shot back. 'I'm worried I'll have some involuntary spasm.'

Ever discreet, Russell wouldn't reveal much about what she had discovered from studying her prize so far. But a few weeks later I phoned Dante Lauretta at the University of Arizona, the principal investigator of the OSIRIS-REx mission, and asked him what he could tell me about the findings so far. And to my surprise, he dropped some real zingers.

When we spoke, scientists had mainly analysed the samples using X-ray diffraction, which tells them the basics about what minerals they contain. This, Lauretta explained, had shown that Bennu is mostly made of clays, specifically a type of mineral called serpentinite. We have that mineral at the bottom of Earth's oceans too, where rocks from the mantle are pushed upwards into the water. The reaction is exothermic, meaning it releases heat, and these rocks are associated with hydrothermal vents on the sea floor – the ones you see in nature documentaries where strange deep-sea creatures squiggle around the warmth of the underwater geyser. On top of that, Lauretta said that the scientists had found a rare type of phosphate mineral coating many of the samples from the asteroid. The only other place such a mineral has been found is in the plumes of water shooting out of Enceladus into space. Put all this together, Lauretta said, and his working hypothesis is that Bennu is a fragment of what was once an ancient ocean world, a fledgling planet covered with water that was destroyed by some massive collision.

Most exciting of all, to my mind, was when Lauretta showed me pictures of what he called 'nanoglobules'. Think of them as the remnants of what would have once been like droplets of oil in water. They are tiny bubbles – or 'protocells', as Lauretta put it. He was not suggesting these were alive, but when you have bubbles like this it creates a separation between the chemical systems inside and without and this is probably a prerequisite for life to start. In other words, Bennu's surface was once the bottom of some warm, extraterrestrial sea where many of the ingredients for life were present. All of this was quite thrilling to hear and I wrote up the story as a minor scoop for *New Scientist*. It made me think again that we really are living through a golden age of science when it comes to asteroids.

It would also be fair to say that of meteoritics more broadly. When I first set out to write about space rocks, I just wanted to answer two straightforward questions: how do you find them, and what do they tell us? But I got a lot more than I bargained for. For instance, the rise and rise of fireball tracking camera networks is, I think, going to be a game-changer. And Jon Larsen's incredible methods for finding urban micrometeorites have opened up a whole new area of science. When it comes to the regular-sized meteorites, the raw numbers alone tell a compelling story. At the time I wrote an early magazine story about meteorites in about 2020, there were about 60,000 specimens in our collections worldwide. Checking the number in early 2024, it had soared to over 73,000. If finds keep accumulating at that rate, we might have 100,000 meteorites by around 2032. That will be quite the milestone.

As we keep hunting, we will inevitably come across ever rarer and stranger ones. There's already hints of this, actually. When I wrote about planetary meteorites earlier in the book, I said that we had only discovered examples from Mars and the moon. But that may no longer be true. In 2023, researchers in France analysed a meteorite found in the Sahara called NWA 13188 and, based on its chemistry, concluded that it had originated on Earth. It was a boomerang meteorite: it was blasted from the ground into space, where it orbited for 100,000 years before plummeting back down. This hypothesis hasn't yet been proven, but in principle 'Earth meteorites' ought to exist and I can't help wondering what kinds of subtle records they might conceal. This story also underlines why studying each and every meteorite is so vital: every so often, we get a shock. That's why I hope that governments will get their acts together and make new laws to guarantee these meteorites are not squirrelled away as collectors' pieces and that their scientific value can always be realised.

On thing's for sure: I won't forget my time walking and talking with meteorite hunters. Before I left the Thorsberg quarry with Birger Schmitz, I selected a palm-sized slab of waste stone and brought it back with me to London. The airport security officers exchanged uncertain glances when they found it in my backpack, but it now sits on my desk, and a half-drunk cup of coffee is resting upon it as I type this. Sometimes I pick it up and wonder at the abstract patterns imprinted on its smooth, cold surface by the shells of ancient sea creatures. It doesn't, of course, contain any fossil meteorites. But I like to keep it close by as I write, a reminder that sometimes you can find the most wonderful things in the most unlikely places.

Acknowledgements

'The heavens declare the glory of God,' wrote the psalmist. 'And the skies proclaim the work of his hands.' As a Christian, the natural world is for me ultimately God's creation. It has been a joy to immerse myself in the world of meteorites and find out how wonderfully they are made. Thank you, God, for your creation – and the fun we can have trying to understand it.

Throughout my career writing about science, I have been amazed at how willingly scientists will answer my emails and take time out of their schedules to explain their work to me. The same goes for this book. It would not have happened without dozens upon dozens of scientists sparing me their time. I am deeply grateful to all of you.

My research for this book took me on several reporting trips: to Manchester and Wold Newton in the UK, and then further afield to Oslo in Norway and Kinnekulle in Sweden. I want to especially thank the people who guided me on those trips. Martin Goff showed me around Wold Cottage on a dark and freezing afternoon. Katie Joy went out of her way to help me from the moment when I first emailed her, and I'm grateful to her, Geoff Evatt and Andy

Smedley for hosting me so kindly in Manchester. I loved spending time with Jon Larsen in Oslo; we had some deep conversations in that white van and I am the richer for it. Thanks also to Birger Schmitz for his candour and kindness in showing me around a very special (and very muddy) mine and introducing me to the glories of Swedish cream and mushroom soup. I also want to give a mention to Jemma Davidson at Arizona State University, who doesn't appear much in these pages, but whose thoughts and insights were repeatedly of great value to me.

Since this is my first book, I want to take this opportunity to thank some of the people who helped me get a foothold in journalism in the first place. A huge thanks to Patrick Walter for taking a chance on me all those years ago and being a gracious editor at *Chemistry World*. To Davide Castelvecchi for being a friendly face at *Nature*. To Mićo Tatalovic and to Anita Makri from the ever-interesting days at SciDev.Net. And to my many brilliant colleagues at *New Scientist*. There are too many of you to name, but a special thanks goes to Dan Cossins and Helen Thomson in the features team. When I say it is 'delightful' to work with you both, you will know it is my highest praise. Most of all, I must thank the inimitable Richard Webb, who gave me a job at *New Scientist* and who – in his own bizarre and wonderful style – has taught me more than anyone else about how to write. He also generously read the whole first draft of this book and gave me brilliant suggestions for how to improve it. Richard is a genius editor and writer and I urge you to read his back catalogue at *New Scientist*: it is a feast for the mind.

My friends have had to endure me talking about this book for years; sometimes with confidence and passion, sometimes with uncertainty and exhaustion. Thank you to Luke and Jo Allen, Ashley and Ellen Haggan, the Lindseys, the Choongs, the Ganpatsinghs, all the family at Christ Church Purley, to the fifty-two crew – especially Nick Funnell – and so many others.

A few years ago, I asked Alex O'Brien if we could have a coffee and talk about books. In that brief meeting, I discovered an unexpectedly kind and enthusiastic colleague, and I'm hugely grateful to Alex for introducing me to Elizabeth Sheinkman, who became my agent. My deepest thanks to you, Elizabeth, for believing in this book from so early on and for backing me so steadfastly. Thank you also to everyone at Oneworld, especially Hannah Haseloff and Sam Carter, my brilliant and tireless editors, for your encouragement and many insightful comments.

My parents-in-law, Audrey and Ed, have proved to be the source of some of my most hilarious anecdotes. The tale of Audrey and the oven trays still gets laughs when I tell it today. As well as being utterly nuts, they have taught me much about what generosity and love is, in ways they probably don't even know themselves. Thank you both.

I struck it lucky with my parents, Heather and David – I'm not sure there are any better ones out there. My mum spent hour upon hour with me as a boy at our kitchen table in Ipswich teaching me to love words and writing. I will always remember those times and be grateful to you, mum, for all your efforts. And, hey, now I've written a book – it must have done some good. I always think my dad exemplifies a form of the common writing tip 'show don't tell'. He isn't one to pontificate or give advice, but he is a steadfast example of how to live a life characterised by thoughtfulness and patience. Thank you also to my brilliant 'Howgebros' Caleb and Seth, whom I love and respect in equal measure.

The last bit is challenging, because even my best-crafted sentences could not express the depth of my love for my children and my beautiful, clever, completely amazing wife Harriet. Thank you for allowing me the space and time to write this book. But more importantly, thank you for everything you give me every day: the smiles, the joy, the dancing in the front room, the games of 'weasel catcher'. The story I care about above all else is the one we will write together.

Notes on Sources

Chapter One: The First Meteorites

Throughout this chapter I draw on Ursula Marvin's detailed review of the early history of meteoritics: U. B. Marvin (1996), 'Ernst Florens Chladni (1756–1827) and the origins of modern meteorite research', *Meteoritics and Planetary Science*, 31 (5), pp. 545–588. Available online at: https://doi.org/mvjx.

Many of my facts about the Wold Cottage meteorite are drawn from: C. T. Pillinger and J. M. Pillinger (1996), 'The Wold Cottage Meteorite: Not just any old ordinary chondrite', *Meteoritics and Planetary Science*, 31 (5), p. 593. Available online at: https://doi.org/mvjz.

On Tutankhamun's dagger: D. Comelli *et al.* (2016), 'The meteoritic origin of Tutankhamun's iron dagger', *Meteoritics and Planetary Science*, 51 (7), pp. 1301–1309. Available online at: https://doi.org/10.1111/maps.12664.

Chapter Two: Modern Meteorite Hunters

My account of Svend Buhl's expeditions is based on both detailed interviews with him and his book: S. Buhl (2019), *For a fistful of rocks: Arid zone meteorite prospection (vol. 1)*, Meteorite Recon, Bad Oldesloe.

Buhl has also posted some of his own memories of his travels on his website, and I draw on some of these. See: https://www.meteorite-recon.com/home/tenere.

NOTES ON SOURCES

The history of the Chinguetti meteorite is documented in: U. B. Marvin (2007), 'Theodore Andre Monod and the lost Fer Dieu meteorite of Chinguetti, Mauritania', *Geological Society, London, Special Publications*, 287, pp. 191–205. Available online at: https://doi.org/d9qmjp.

Robert Warren's analysis of where he thinks the fabled Chinguetti meteorite lies: R. Warren, S. Warren and E. Protopapa (2024) 'New evidence on the lost giant Chinguetti meteorite', preprint posted to the arXiv server. Available online at: https://arxiv.org/abs/2402.14150.

***'I know meteorites, I know danger, and I know how to have a good time.'* This quote comes from Robert Ward's video:** 'Space Cowboy take two'. Available online at: https://www.youtube.com/watch?v=mk_UbWK4ylw.

Some of my facts about Warren and Farmer's time in jail are taken from: R. Highfield (2011) 'Meteorite hunter: My two months in an Omani jail', *New Scientist*, 29 June. Available online at: https://www.newscientist.com/article/mg21128190-200-meteorite-hunter-my-two-months-in-an-omani-jail/.

***'I came close to sinking a shovel into one guy's skull,'* Farmer told journalists reporting for *Science* magazine. This quote comes from a wonderful magazine feature that provides a detailed account of the hunt for fragments of Aguas Zarcas, and which features additional reporting from the Costa Rica-based journalist Andrea Solano Benavides:** J. Sokol (2020), 'Lucky strike', *Science*, 369 (6505), pp. 750–765. Available online at: https://doi.org/ksjg.

Amino acids in the Aguas Zarcas meteorite: D. P. Glavin *et al.* (2021), 'Extraterrestrial amino acids and L-enantiomeric excesses in the CM2 carbonaceous chondrites Aguas Zarcas and Murchison', *Meteoritics and Planetary Science*, 56 (1), pp. 148–173. Available online at: https://doi.org/kx5h.

Information on the requirements for having a meteorite officially certified is taken from: The Meteoritical Society, 'Guidelines on meteorite nomenclature'. Available online at: https://www.lpi.usra.edu/meteor/docs/nc-guidelines.pdf.

***One man told of how he would ride his motorcycle into the desert and pick up as many likely-looking rocks as he could*:** K. R. Rosen (2017) 'The Meteorite Hunters: the Booming Trade in Space Rocks', *Financial Times*, 16 November. Available online at: https://www.ft.com/content/a8e84944-c988-11e7-aa33-c63fdc9b8c6c.

The episode involving Steve Arnold finding the meteorite in Normandy was reported in the French press: J-B. Jacquin (2023) 'The Chase Between Meteorite Hunters and the French Natural History Museum', *Le Monde*, 3 June. Available online at: https://www.lemonde.fr/en/france/article/2023/06/03/the-chase-between-meteorite-hunters-and-the-french-natural-history-museum_6028935_7.html.

Steve Arnold's video from a beach in Normandy: S. Arnold (2023), Video posted to Facebook, 19 February. Available online at: https://www.facebook.com/1242070021/videos/764763591747515/.

After all, another meteorite fell in France in September 2023, making that two falls in the country in a little over nine months: A. Steinhauser (2023), 'Une nouvelle meteorite français!', *Vigie Ciel*, 13 September. Available online at: https://www.vigie-ciel.org/2023/09/13/une-nouvelle-meteorite-francaise/.

Chapter Three: Tracking Fireballs

On the nature of the Chelyabinsk bolide: O. Popova (2013), 'Chelyabinsk airburst, damage assessment, meteorite recovery, and characterization', *Science*, 342 (6162), pp. 1,069–1,073. Available online at: https://doi.org/f5h73v.

On the Harvard Meteor Programme: L. G. Jacchia and F. L. Whipple (1956), 'The Harvard photographic meteor programme', *Vistas in Astronomy*, 2 (1), pp. 982–994. Available online at: https://doi.org/cds6fq.

***A licence plate that read* 'COMETS':** K. Chang, 'Fred L. Whipple Is Dead; Expert on Comets Was 97', *New York Times*, 31 August 2004. Available online at: https://www.nytimes.com/2004/08/31/us/fred-l-whipple-is-dead-expert-on-comets-was-97.html.

Details about Zdeněk Ceplecha's early life and work: Based on an interview with Pavel Spurný and Z. Ceplecha (1960) 'Multiple fall of Pribram meteorite photographed. 1. Double-station photographs of the fireball and their relations to the found meteorites', *Bulletin of the Astronomical Institute of Czechoslovakia*, 12, pp. 21–47.

Modelling of the Pribram meteorite's orbit in 2008: M. Gritsevich (2008), 'The Pribram, Lost City, Innisfree, and Neuschwanstein falls: An analysis of the atmospheric trajectories', *Solar System Research*, 42, pp. 372–390. Available online at: https://doi.org/bzcx84.

The orbit of the Neuschwanstein meteorite: P. Spuny *et al.* (2003), 'Photographic observations of Neuschwanstein, a second meteorite from the orbit of the Příbram chondrite', *Nature*, 423, pp. 151–153. Available online at: https://doi.org/d26nj3.

A list of meteorite falls with photographed orbits is maintained online by Matthias Meier at the Natural History Museum of St. Gallen, Switzerland: Available online at: https://www.meteoriteorbits.info/.

For information on the Orgueil meteorite, I drew on: M. Gounelle and M. E. Zolensky (2014), 'The Orgueil meteorite: 150 years of history', *Meteoritics and Planetary Science*, 49 (10), pp. 1,769–1,794. Available online at: https://doi.org/f6n6bw.

Calculating the orbit of the Orgueil meteorite: M. Gounelle, P. Spurný and P. A. Bland (2006), 'The orbit and atmospheric trajectory of the Orgueil meteorite from historical records', *Meteoritics and Planetary Science*, 41 (1), pp. 135–150. Available online at: https://doi.org/fr66jz.

The Bunburra Rockhole meteorite: P. Bland *et al.* (2009), 'An anomalous basaltic meteorite from the innermost main belt', *Science*, 325 (5947), pp. 1,525–1,527. Available online at: https://doi.org/c8nw8n.

On field tests of a drone for spotting meteorites: S. Anderson *et al.* (2020), 'Machine learning for semi-automated meteorite recovery', *Meteoritics and Planetary Science*, 55 (11), pp. 2,461–2,471. Available online at: https://doi.org/mvj4.

Only a handful of meteorites are seen to fall and then be actually recovered around the world each year – in 2016, for example, there were only eleven. This is taken from: C. Smith, S. Russell and N. Almeida (2019), *Meteorites: The Story of Our Solar System*, Firefly Books, p. 11.

'...my original thought was: has someone been driving around the Cotswolds lobbing lumps of coal into people's gardens?': J. Amos (2021), *A fireball, a driveway and a priceless meteorite*, BBC News. Available online at: https://www.bbc.co.uk/news/science-environment-56337876 (accessed 8 January 2022).

Mira Ihasz found a fragment of the Winchcombe meteorite at exactly 09:39 in the morning. This is based on social media accounts available online at: https://twitter.com/mirabii/status/1500442806376091651.

A video of Ihasz and Daly celebrating riotously in the aftermath of the discovery is available online at: https://twitter.com/aineclareob/status/1500475515509084164.

Chapter Four: The Mystery of the Missing Meteorites

My account of William Cassidy's adventures draws on interviews with those who worked with him at the time and on his brilliantly written memoir: W. Cassidy (2003), *Meteorites, Ice and Antarctica*, Cambridge University Press.

Cassidy died in 2020 aged ninety-two. An obituary was published by The Meteoritical Society and is available online at: https://meteoritical.org/news/william-cassidy-1928-2020.

My writing about the 'missing iron meteorites of Antarctica' in this chapter focuses on Geoff Evatt, Katie Joy and Andrew Smedley for narrative purposes. However, they were the leaders of a much larger team in the search for Antarctica's missing meteorites. This included David Abrahams, Jane MacArthur, Antony Peyton, Liam Marsh, John Wilson, Wouter Van Verre, Romain Tartèse and John Davidson. The project was also supported by Mike Rose and Laura Gerrish at the British Antarctic Survey. And the polar field guides included Julie Baum, Taff Raymond and Rob Taylor. To mention everyone involved in this project would make for a much longer list, which space precludes.

Geoff Evatt's rewilding project is called Sunart Fields: https://www.sunartfields.com/.

The experiments looking at how quickly meteorites sink into ice: G. W. Evatt *et al.* (2016), 'A potential hidden layer of meteorites below the ice surface of Antarctica', *Nature Communications*, 7, article number 10679. Available online at: https://doi:10.1038/ncomms10679.

The effect of climate change on Antarctica meteorites: V. Tollenaar *et al.*, (2024), 'Antarctic meteorites threatened by climate warming', *Nature Climate Change*, 14, pp. 340–344. Available online at: https://doi.org/mvbw.

17,600 stones weighing more than fifty grams each year: G. W. Evatt, A. R. D. Smedley, K. H. Joy, L. Hunter, W. H. Tey, I. D. Abrahams and L. Gerrish (2020), 'The spatial flux of Earth's meteorite falls found via

Antarctic data', *Geology*, 48 (7), pp. 683–687. Available online at: https://doi.org/10.1130/G46733.1.

The meteorites were expected to be only five to ten centimetres under the ice: A. R. Smedley, G. W. Evatt, A. Mallinson and E. Harvey (2020), 'Solar radiative transfer in Antarctic blue ice: spectral considerations, subsurface enhancement, and inclusions', *The Cryosphere*, 14 (3), pp. 789–809. Available online at: https://doi.org/mvj7.

The number of meteorites recovered from Antarctica is more than 47,000. Based on a search of the Meteoritical Bulletin Database in April 2023. Available online at: https://www.lpi.usra.edu/meteor/metbull.php.

Chapter Five: On the Rooftops

My writing about Jon Larsen's life and work is based on extensive interviews with him and on his book: J. Larsen (2021), *Star Hunter*, Arthouse DGB.

Nininger was convinced these were real micrometeorites: H. H. Nininger (1941), 'Collecting small meteoritic particles', *Contributions of the Society for Research on Meteorites*, 2 (7), pp. 258–260. Available online at: https://doi.org/mvj8.

'The notion that micrometeorites could be recovered from urban environments was considered an urban myth…' **This quote is taken from:** M. Suttle, T. Hasse and T. Hecht (2021), 'Evaluating urban micrometeorites as a research resource – A large population collected from a single rooftop', *Meteoritics and Planetary Science*, 56 (8), pp. 1,531–1,555. Available online at: https://doi.org/mvj9.

Matthew Genge's system for classifying micrometeorites: M. J. Genge *et al.* (2008), 'The classification of micrometeorites', *Meteoritics and Planetary Science*, 43 (3), pp. 497–515. Available online at: https://doi.org/d3nrhn.

Descriptions of the different kinds of micrometeorites that Larsen identified (together with beautiful photographs of them) can be found in: J. Larsen (2019), *On the Trail of Stardust*, Voyeur Press.

Discovery of the first urban micrometeorites: M. J. Genge, J. Larsen *et al.* (2017), 'An urban collection of modern-day large micrometeorites: Evidence for variations in the extraterrestrial dust flux through the

Quaternary', *Geology*, 45 (2), pp. 119–122. Available online at: https://doi.org/bvm2.

Details of Scott Peterson's micrometeorite work are available online at: https://micro-meteorites.com/.

Martin Suttle's study of the flux of cosmic dust falling on Earth: M. D. Suttle and L. Folco (2020), 'The extraterrestrial dust flux: Size distribution and mass contribution estimates inferred from the Transantarctic Mountains (TAM) micrometeorite collection', *JGR Planets*, 125 (2). Available online at: https://doi.org/mvkc.

On the spherules purportedly from an interstellar meteorite: A. Loeb *et al.* (2023), 'Discovery of spherules of likely extrasolar composition in the Pacific Ocean site of the CNEOS 2014-01-08 (IM1) bolide'. Available online at: https://arxiv.org/abs/2308.15623.

Chapter Six: An Unexpected History of the Solar System

The planet-forming discs observed by the ALMA telescopes were reported in: R. Parker *et al.* (2022), 'Taxonomy of protoplanetary discs observed with ALMA', *Monthly Notices of the Royal Astronomical Society*, 511 (2), pp. 2,453–2,490. Available online at: https://doi.org/kvq2.

The Nice model was published as a trio of papers, the first of which was: R. Gomes *et al.* (2005), 'Origin of the cataclysmic Late Heavy Bombardment period of the terrestrial planets', *Nature*, 435, pp. 466–496. Available online at: https://doi.org/bkd9zf.

The Grand Tack model: K. Walsh *et al.* (2011), 'A low mass for Mars from Jupiter's early gas-driven migration', *Nature*, 475, pp. 206–209. Available online at: https://doi.org/dsq69j.

On the patterns Paul Warren spotted in isotopic data: P. H. Warren (2011), 'Stable-isotopic anomalies and the accretionary assemblage of the Earth and Mars: A subordinate role for carbonaceous chondrites', *Earth and Planetary Science Letters*, 311 (1–2), pp. 93–100. Available online at: https://doi.org/dczjjg.

Thorsten Kleine's extension of Warren's work to include molybdenum isotopes: G. Budde *et al.* (2016), 'Molybdenum isotopic evidence for the origin of chondrules and a distinct genetic heritage of carbonaceous and non-carbonaceous meteorites', *Earth and Planetary Science Letters*, 454 (15), pp. 293–303. Available online at: https://doi.org/f89sv8.

Using the great dichotomy to date the formation of Jupiter: T. S. Kruijer *et al.* (2017), 'Age of Jupiter inferred from the distinct genetics and formation times of meteorites', *Proceedings of the National Academy of Sciences*, 114 (26) pp. 6,712–6,716. Available online at: https://doi.org/gpgjq4.

Ben Weiss's work on the vanishing of the solar nebula's magnetic field: H. Wang, B. P. Weiss *et al.* (2017), 'Lifetime of the solar nebula constrained by meteorite paleomagnetism', *Science*, 355 (6,325), pp. 623–627. Available online at: https://doi.org/f9qwfn.

Weiss's work constraining the growth of Jupiter: B. P. Weiss and W. F. Bottke (2012) 'What can meteorites tell us about the formation of Jupiter?', *AGU Advances*, 2 (2), e2020AV000376. Available online at: https://doi.org/kx8t.

How metal detectors can erase the vestigial magnetism of meteorites: F. Vervelidou, B. P. Weiss and F. Lagroix (2023), 'Hand magnets and the destruction of ancient meteorite magnetism', *JGR Planets*, 128 (4), e2022JE007464. Available online at: https://doi.org/gr4pvz.

David Nesvorný's addition of an extra gas giant to the Nice model: D. Nesvorný (2011), 'Young solar system's fifth giant planet?', *The Astrophysical Journal Letters*, 742 (L22). Available online at: https://doi.org/dkqmtz.

Chapter Seven: The Origins of Our Oceans

Angus Barbieri's case is described in: W. K. Stuart and L. W. Fleming (1973), 'Features of a successful therapeutic fast of 382 days' duration', *Postgraduate Medical Journal*, 49, pp. 203–209. Available online at: https://doi.org/c3cc35.

The facts about the amount of water on Earth are adapted from: U.S. Geological Survey Water Science School (2019), 'How much water is there on Earth?' Available online at: https://www.usgs.gov/special-topics/water-science-school/science/how-much-water-there-earth.

A summary of the scientific story around deuterium–hydrogen ratios in materials from across the solar system and what they mean is given in:

K. Meech and S. Raymond (2020), 'Origin of Earth's water: Source and Constraints' in V. S. Meadows, V. N. Arney, B. E. Schmidt and B. J. Des Marias (eds), *Planetary Astrobiology*, University of Arizona Press.

On the reasons why there seems to have been more deuterium in the outer edges of the protoplanetary disc: L. I. Cleeves *et al.* (2016), 'Exploring the origins of deuterium enrichments in solar system nebular organics', *The Astrophysical Journal*, 819 (13). Available online at: https://doi.org/mbds.

The value for the sun's deuterium–hydrogen ratio was reported in: J. Geiss and G. Gloeckler (1998), 'Abundances of deuterium and helium-3 in the protosolar cloud', *Space Science Reviews*, 84, pp. 239–250. Available online at: https://doi.org/c786dm.

Concerns around the cyanogen gas in the tail of Halley's comet were aired in: Unknown author (1910), 'COMET'S POISONOUS TAIL: Yerkes Observatory Finds Cyanogen in Spectrum of Halley's Comet', *New York Times*, 8 February, p.1. Available online at: https://www.nytimes.com/1910/02/08/archives/comets-poisonous-tail-yerkes-observatory-finds-cyanogen-in-spectrum.html.

Comet Hale-Bopp's D/H ratio was reported in: R. Meier *et al.* (1998), 'A determination of the HDO/H_2O ratio in Comet C/1995 O1 (Hale-Bopp)', *Science*, 279 (5352), pp. 842–844. Available online at: https://doi.org/fbm4vc.

Comet Hartley 2's D/H ratio was reported in: P. Hartogh *et al.* (2011), 'Ocean-like water in the Jupiter-family comet 103P/Hartley 2', *Nature*, 478, pp. 218–220. Available online at: https://doi.org/fswd6c.

Comet 67P's D/H ratio was reported in: K. Altwegg *et al.* (2014), '67P/Churyumov-Gerasimenko, a Jupiter family comet with a high D/H ratio', *Science*, 347 (6220). Available online at: https://doi.org/f6zs8b.

The idea that the carbonaceous chondrites could have delivered Earth's water: F. Robert (2003), 'The D/H ratio in chondrites', *Space Science Reviews*, 106, pp. 87–101. Available online at: https://doi.org/bznd9x.

Lydia Hallis's measurements of Earth's primordial water: L. J. Hallis *et al.* (2015), 'Evidence for primordial water in Earth's deep mantle', *Science*, 350 (6262), pp. 795–797. Available online at: https://doi.org/875.

John Bradley's experiment simulating water generation on powdered rock: C. Zhu *et al.* (2018), 'Untangling the formation and liberation of water in the lunar regolith', *Proceedings of the National Academy of Sciences*, 116 (23), pp. 11,165–11,170. Available online at: https://doi.org/ggw3d6.

Luke Daly's work looking at the amount of moisture contained in grains from the asteroid Itokawa: L. Daly *et al.* (2021), 'Solar wind contributions to Earth's oceans', *Nature Astronomy*, 5, pp. 1,275–1,285. Available online at: https://doi.org/k8vw.

In 2017, there was a Royal Society Theo Murphy meeting to discuss the origins of water in the solar system. (Incidentally, this is the meeting at which I first became interested in meteorites myself.) A write-up of the discussions is introduced here: S. Russell, C. Ballentine and M. Grady (2017), 'The origin, history and role of water in the evolution of the inner Solar System', *Philosophical Transactions of the Royal Society A*, 375 (2094). Available online at: https://doi.org/gd57tm.

The story of Zag and Monahans meteorites and Mike Zolensky's analysis of the fluid inclusions inside them: Zolensky *et al.* (2017), 'The search for and analysis of direct samples of early Solar System aqueous fluids', *Philosophical Transactions of the Royal Society A*, 375 (2094). Available online at: https://doi.org/ktpv.

Chapter Eight: Space Fossils

Birger Schmitz's letter to Alvarez: B. Schmitz (1985), 'Metal precipitation in the Cretaceous-Tertiary boundary clay at Stevns Klint, Denmark', *Geochimica et Cosmochimica Acta*, 49 (11), pp. 2,361–2,370. Available online at: https://doi.org/b75jch.

The fossil meteorite found in Jämtland in the 1950s: C. Alwmark and B. Schmitz (2009), 'The origin of the Brunflo fossil meteorite and extraterrestrial chromite in mid-Ordovician limestone from the Gärde quarry (Jämtland, central Sweden)', *Meteoritics & Planetary Science*, 44 (1), pp. 95–106. Available online at: https://doi.org/dzb65d.

Frank Kyte's claims of a fossil meteorite: F. T. Kyte (1998), 'A meteorite from the Cretaceous/Tertiary boundary', *Nature*, 396, pp. 237–239. Available online at: https://doi.org/fqkxfx.

Schmitz's initial calculation of the flux of meteorites 480 million years ago: B. Schmitz, M. Tassinari and B. Peucker-Ehrenbrink (2001), 'A rain of ordinary chondritic meteorites in the early Ordovician', *Earth and*

Planetary Science Letters, 194 (1–2), pp. 1–15. Available online at: https://doi.org/cswk7k.

Revised estimates of that flux based on data that included grains of chromite: B. Schmitz, T. Håggström and M. Tassinari (2003), 'Sediment-dispersed extraterrestrial chromite traces a major asteroid disruption event', *Science* (300) 5621, pp. 961–964. Available online at: https://doi.org/cmwmp6.

Details of Schmitz's collaboration with Rainer Wieler (see pp. 81–87): R. Wieler (2023), 'A journey in noble gas cosmochemistry and geochemistry', *Geochemical Perspectives*, 12 (1), pp. 1–178. Available online at: https://doi.org/mvkf.

The exposure age of the chromite increased neatly and gradually depending on the age of the rock it was found in: P. R. Heck *et al.* (2004), 'Fast delivery of meteorites to Earth after a major asteroid collision', *Nature*, 430, pp. 323–325. Available online at: https://doi.org/10.1038/nature02736.

Susan Taylor and Donald Brownlee's study on ancient meteorites: S. Taylor and D. E. Brownlee (1991), 'Cosmic spherules in the geologic record', *Meteoritics and Planetary Science*, 26 (3), pp. 203–211. Available online at: https://doi.org/ks6s.

Andy Tomkins' fossil iron micrometeorites: A. G. Tomkins *et al.* (2016), 'Ancient micrometeorites suggestive of an oxygen-rich Archaean upper atmosphere', *Nature*, 533, pp. 235–238. Available online at: https://doi.org/bhgs.

Rebecca Payne's inferences about the composition of the ancient atmosphere based on the oxidation of fossil meteorites: R. C. Payne, J. Kasting and B. Brownlee (2020), 'Oxidized micrometeorites suggest either high pCO2 or low pN2 during the Neoarchean', *Proceedings of the National Academy of Sciences*, 117 (3), pp. 1,360–1,366. Available online at: https://doi.org/gk3xkh.

The results that backed up Payne's study: O. R. Lehmer *et al.* (2020), 'Atmospheric CO_2 levels from 2.7 billion years ago inferred from micrometeorite oxidation', *Science Advances*, 6 (4). Available online at: https://doi.org/ghwcqn.

Martin Suttle's micrometeorites from the Cretaceous period are described in: M. D. Suttle and M. Genge (2017), 'Diagenetically altered fossil micrometeorites suggest cosmic dust is common in the geological record', *Earth and Planetary Science Letters*, 476, pp. 132–142. Available online at: https://doi.org/gb2jrs.

Suttle analysed a bunch of samples from Italy that were formed about 8 million years ago, during the Miocene period: M. Suttle *et al.* (2023), 'Fossil micrometeorites from Monte dei Corvi: Searching for dust from the Veritas asteroid family and the utility of micrometeorites as a palaeoclimate proxy', *Geochimica and Cosmochimica Acta*, 355, pp. 75–88. Available online at: https://doi.org/ktdw.

Details of Birger Schmitz's EU-funded project to dissolve industrial amounts of rock can be found online at: https://cordis.europa.eu/project/id/291300.

This was enough to reconstruct the flux of L-chondrites over tens of millions of years: B. Schmitz *et al.* (2019), 'An extraterrestrial trigger for the mid-Ordovician ice age: Dust from the breakup of the L-chondrite parent body', *Science Advances*, 5 (9). Available online at: https://doi.org/gjdkxr.

Schmitz's study of cosmic dust flux over the past 500 million years: F. Terfelt and B. Schmitz (2021) 'Asteroid break-ups and meteorite delivery to Earth the past 500 million years', *Proceedings of the National Academy of Sciences*, 118 (24), e2020977118. Available online at: https://doi.org/gkx8t4.

Chapter Nine: Where Meteorites Come From

The postcards impregnated with the smell of comet 67P were created in 2016 and written about by several journalists at the time, including my colleague Jacob Aron at *New Scientist*. The cards had been all but used up; I thank Geraint Jones at University College London for kindly digging out a few spares and sending them to me.

On the LDEF mission: M. Zolensky (2021), 'The Long Duration Exposure Facility – a forgotten bridge between Apollo and Stardust', *Meteoritics and Planetary Science*, 56 (5), pp. 900–910. Available online at: https://doi.org/gjxjsn.

On the Genesis mission: D. S. Burnett (2013), 'The Genesis solar wind sample return mission: Past, present, and future', *Meteoritics and Planetary Science*, 48 (12), pp. 2,351–2,370. Available online at: https://doi.org/f5m5s7.

NOTES ON SOURCES

On the Orbital Debris Collector experiment: F. Hörz *et al.* (2000), 'Impact features and projectile residues in aerogel exposed on Mir', *Icarus* 147 (2), pp. 559–579. Available online at: https://doi.org/c8mcvj.

Details of NASA's experiments with the tomato seeds flown into space can be found in: 'Attack of the killer space tomatoes? Not!' (1992). Available online at https://www.nasa.gov/home/hqnews/1992/92-049.txt.

The idea of photochemical self-shielding was first described in: R. N. Clayton, 'Self-shielding in the solar nebula' (2002), *Nature*, 415, pp. 860–861. Available online at: https://doi.org/dsmhhb.

In 2006, the journal *Science* devoted an issue to Hayabusa's discoveries about the asteroid Itokawa. It is available online at: https://www.science.org/toc/science/312/5778.

Erik Asphaug's quote about Itokawa is taken from: E. Asphaug (2006), 'Adventures in near-earth object exploration', *Science*, 312 (5778), pp. 1,328–1,329. Available online at: https://doi.org/bhhfcb.

Ryugu turned out to be a close chemical match for CI-chondrites: R. C. Greenwood *et al.* (2023), 'Oxygen isotope evidence from Ryugu samples for early water delivery to Earth by CI chondrites', *Nature Astronomy*, 7, pp. 29–38. Available online at: https://doi.org/mvkj.

They found carbon dioxide buried inside its crystals: T. Nakamura *et al.* (2022), 'Formation and evolution of carbonaceous asteroid Ryugu: Direct evidence from returned samples', *Science*, 379 (6634). Available online at: https://www.doi.org/10.1126/science.abn8671.

A special issue of Science dedicated to the results from Hayabusa 2 is available online at: https://www.science.org/toc/science/379/6634.

Details of the Psyche mission can be found online at: https://www.jpl.nasa.gov/missions/psyche.

Information about the CAESAR mission is available online at: https://caesar.cornell.edu/mission/.

Epilogue

My news story based on the interview with Dante Lauretta: J. Howgego (2024), 'Asteroid sampled by NASA may once have been part of an ocean world', *New Scientist*, 6 February. Available online at: https://www.newscientist.com/article/2415791-asteroid-sampled-by-nasa-may-once-have-been-part-of-an-ocean-world/.

On the 'boomerang meteorite', NWA 13188: A. Wilkins (2023), 'Meteorite left Earth then landed back down after round trip to space', *New Scientist*, 11 July. Available online at: https://www.newscientist.com/article/2381928-meteorite-left-earth-then-landed-back-down-after-round-trip-to-space/.